Gender and Climate Change:
An Introduction

Gender and Climate Change:
An Introduction

Edited by Irene Dankelman

publishing for a sustainable future

London • Sterling, VA

First published in 2010 by Earthscan

Revenues of this book will be donated to the Jake Waller-Hunter Initiative to strengthen local women's leadership in climate change policy and action; see www.bothends.org/index/php?page=2_2 and Box 9.2 (p 228).

Earthscan Ltd, Dunstan House, 14a St Cross Street, London EC1N 8XA, UK
Earthscan LLC, 1616 P Street, NW, Washington, DC 20036, USA

Earthscan publishes in association with the International Institute for Environment and Development

For more information on Earthscan publications, see www.earthscan.co.uk or write to earthinfo@earthscan.co.uk

ISBN: 978-1-84407-864-6 hardback
ISBN: 978-1-84407-865-3 paperback

Typeset by JS Typesetting Ltd, Porthcawl, Mid Glamorgan
Cover design by Susanne Harris

A catalogue record for this book is available from the British Library

Library of Congress Cataloging-in-Publication Data
Gender and climate change : an introduction / edited by Irene Dankelman.
    p. cm.
  Includes bibliographical references and index.
  ISBN 978-1-84407-864-6 (hardback) — ISBN 978-1-84407-865-3 (pbk.)
  1. Climatic changes—Research. 2. Climatic changes—Sex differences. 3. Sex role—Research. I. Dankelman, Irene.
  QC903.G47 2010
  304.2'5—dc22

                              2010027613

At Earthscan we strive to minimize our environmental impacts and carbon footprint through reducing waste, recycling and offsetting our CO_2 emissions, including those created through publication of this book. For more details of our environmental policy, see www.earthscan.co.uk.

Printed and bound in the UK by
CPI Antony Rowe, Chippenham, Wiltshire
The paper used is FSC certified.

Contents

Part II Realities on the Ground

Part III Strategies and Action

List of Figures, Tables and Boxes

Figures

Tables

Boxes

List of Contributors

Lorena Aguilar is senior gender advisor to the International Union for Conservation of Nature (IUCN). She is an international advisor for numerous organizations, governments and universities on topics related to water, environmental health, and gender and community participation. With a master's degree in anthropology, Aguilar, who majored in cultural ecology at the University of Kansas, has worked for nine years in the field of development and in the design of public policy projects in Central America. For the past decade she has been actively engaged in the incorporation of social and gender aspects into the use and conservation of natural resources worldwide. She has published widely about gender and environment, environmental health and public policy involving equity issues.

Emma R. M. Archer is principal researcher in climate change at the Council for Scientific and Industrial Research (CSIR) Natural Resources and the Environment in South Africa. Archer holds a PhD and was trained at the University of Cape Town and at Clark University in the US. She has worked on climate change for more than a decade with a particular focus on climate change and managed ecosystems. Archer and her team are working throughout southern Africa.

Bernhard Barth is a member of the Cities in Climate Change project team of UN-HABITAT. He is based in Nairobi. He is the focal point for the pilot initiative in Sorsogon. Barth is also the gender focal point of UN-HABITAT's Training and Capacity Building Branch and as such oversaw the development of the *Gender in Local Government – A Sourcebook for Trainers* tool. He conducts gender training for urban practitioners.

Eleanor Blomstrom is the sustainable development and climate change coordinator for the Women's Environment & Development Organization (WEDO). At WEDO, Blomstrom focuses on the gender and international aspects of climate change. Her work involves research, global-level advocacy, primarily at UN fora, and programmatic work with national partners around the world. She holds a master's degree in international affairs in urban and environmental policy and a bachelor's degree in environmental sciences.

Sabine Bock is director for Germany and Energy and Climate Change coordinator of Women in Europe for a Common Future (WECF). Before she joined WECF in 2006, she worked for Greenpeace New Zealand and Green City, Germany, where she was involved in energy and climate issues. In her work for WECF she follows the United Nations Framework Convention on Climate Change (UNFCCC), Commission on Sustainable Development (CSD), United Nations Environment Programme (UNEP) and Environment for Europe (EfE) processes and supports the implementation of sustainable energy solutions locally with a focus on gender and human rights aspects.

Ruth Bond is the chair of the National Federation of Women's Institutes (NFWI) across England and Wales. Previously, she was the national Public Affairs Committee chairman. She is also a board member of the UK's largest climate change coalition Stop Climate Chaos. She was instrumental in establishing Cambridge Federation's sustainable headquarters and is a long-standing champion of green issues within the NFWI.

Bharati Chaturvedi, who has a master's degree in history and a master's in international public policy, is the founder and director of the Chintan Environmental Research and Action Group. The objective of her partnership with the urban poor was to find ways to foster environmental justice. Chaturvedi also writes on environmental and development issues, for example, the weekly column Earthwatch in the *Hindustan Times*. She talks about these issues widely, in India and internationally, as part of her advocacy and outreach and has co-produced three films: *Waste, 60 Kilos* and *Citizens at Risk*. She has been awarded a LEAD Fellowship, is a Fellow at the Synergos Institute and recently received the Knowledge for the World Award (Johns Hopkins Alumni Association).

Emily Cleevely is Research and Campaigns officer at the NFWI in the UK and a member of the Public Affairs team at the NFWI, responsible for running campaigns on the environment and climate change. She ran the successful Women and Climate Change campaign in 2009 and helped to achieve the NFWI's biggest turnout at a national climate change demonstration The Wave in December 2009.

Thais Corral has been implementing global and local initiatives geared to strengthen the role of women in the sustainable development platform for the last two decades. She is the founder of two non profit organizations in Brazil, REDEH (Human Development Network) and CEMINA, that is engaged in communication for and by women. In 1992, she was the co-organizer of the Planeta Femea (Women's Tent) at the Earth Summit in Rio De Janeiro, and she is one of the co-founders of WEDO. Corral was the capacity building director of the SouthSouthNorth collaborative that leapfrogged projects on mitigation and adaptation to climate change with poverty reduction in six countries of Africa, Latin America and Asia. One of these projects, which she directly coordinated, Pintadas Solar, won the 2008 Seed Award and was recognized as a best practice by the United Nations Human Settlements Programme (UN-HABITAT) Dubai Award.

Tracy Cull undertook postgraduate research on the mainstreaming of gender in local government structures in the new South Africa. She has a particular interest in the pedagogy around geographical training and education, with more than 13 years of experience as a lecturer at the University of the Witwatersrand, Johannesburg. More recently she has been consulting to public and private sector organizations, designing, and running specialized training courses around various environment and development issues.

Irene Dankelman is an ecologist by background. She has worked for more than 30 years in the area of environment and sustainable development for national and international organizations, academia and the UN. She specialized in the area of gender and environment and has published widely on that subject. Presently, Dankelman is lecturer in sustainable development at the Radboud University in Nijmegen (Netherlands), and a consultant at IRDANA advice. She is member of (advisory) boards of several international women-environment-development organizations, including WEDO and WECF, and of the Joke Waller-Hunter Initiative of Both ENDs (Environment and Development Service for non-governmental organizations).

Gero Fedtke is coordinator of the Empowerment and Local Action (ELA) programme at WECF. In the ELA programme, 30 partner organizations in the EECCA (Eastern Europe Caucasus and Central Asia) region and Afghanistan implement projects on sustainable rural development with a focus on sanitation and energy issues. Fedtke studied Eastern European and Soviet History, Slavic Philology, Political Science and Central Asian Studies at Cologne, Bonn and Volgograd Universities. Before joining WECF in 2005, he taught Russian History at Bochum University, Germany.

Sascha Gabizon has been the executive director of WECF since 1996. She obtained a master's in business, worked for an international consultancy in Spain and then joined the Wuppertal Institute for Climate, Energy and Environment, in Germany, where she was the co-founder of the Wuppertal Institute's 'Frauen Wissen' (Women's Scientists). In 1994, Gabizon joined WECF to prepare its contribution to the Fourth World Women's Conference in Beijing (1995). She developed WECF's network and its activities in the EU, Eastern Europe and Central Asia, and its secretariats in the Netherlands, Germany and France. Gabizon has long-term project experience in Armenia, Bulgaria, Georgia, Hungary, Kyrgyzstan, Moldova, Romania, Russia, Poland, Ukraine and Uzbekistan. She has published a large number of case studies and articles.

Shana Griffin is a black feminist, social justice activist, researcher and mother. She is co-founder of the Women's Health and Justice Initiative (WHJI), where she currently serves as research and advocacy director. Shana holds a MA in sociology and has more than 15 years experience organizing in low-income and working class communities of colour on critical issues of gender, racial and economic equity.

Rachel Harris, who was born and raised in New Orleans, now works on gender and climate change issues at WEDO in New York City. While at WEDO, she has been particularly concerned about mobilizing US women to be a greater part of the climate change debate both in the US and globally. Before coming to WEDO, Harris earned her MA in climate and society from Columbia University, focusing on the impacts of climate variability on health and vulnerable communities. She has previously worked on these issues at the International Research Institute for Climate and Society and the World Resources Institute.

Willy Jansen is professor of gender studies and director of the Institute for Gender Studies at the Radboud University Nijmegen, the Netherlands. She obtained her PhD in anthropology in 1987 from Radboud University Nijmegen, on a study of gender and marginalization in Algeria. She has been senior research fellow at Yarmouk University in Jordan, and professor of anthropology at University of Amsterdam. In 2009 she was elected as member of the Royal Netherlands Academy of Arts and Sciences. Her publications reflect three lines of research: A first line deals with the history of women's education in the Middle East, with Jordan as a case study. In the second research line, a variety of cultural expressions of gender are analysed. A third research line deals with the anthropology of religion, in particular pilgrimage as lived religion. On all these aspects, Jansen has published widely.

Gerd Johnsson-Latham works as deputy director at the Department for Development Policies within the Swedish Ministry of Foreign Affairs. She is a gender advisor and has written several publications on gender and poverty, gender and violence and on gender and sustainable development. Her study 'Gender equality as a prerequisite to sustainable development' was presented at the UN Commission for Sustainable Development and at the High Level Panel on Climate Change, in 2007 in New York, and at the UNEP Ministerial Gender Panel in Nairobi, February 2009.

Janet Kabeberi-Macharia is the UNEP senior gender adviser based at the UNEP headquarters in Nairobi. She holds a PhD in law and her career spans academia, non-governmental organizations (NGOs), and international development organizations with more than 15 years' experience in research, training, implementation, monitoring and evaluation of rights-based programmes. In her work in UNEP, she has focused on building internal staff capacity on gender mainstreaming, developing different methodologies for mainstreaming gender into environmental management programmes. She has extensively published books and articles on gender issues, children's rights, women's human rights, law and development, and environmental law.

Rehana Bibi Khilji belongs to an indigenous tribe of Baluchistan, Pakistan, an area where it was unusual for girls to attend school. After finishing university education with a master's in political science, she started working with a local NGO on the revival of indigenous irrigation systems and social mobilization. In 1998, she and three other women started the volunteer platform HOPE, working with Afghan refugees and host communities on women's education, social protection and capacity building in environmental health. In 2008 she joined HOPE full-time as a team leader, and is responsible for the management of overall programmes and operations on livelihood security, environment, disaster risk reduction and gender equality and human rights.

Prabha Khosla is an urban planner. She works on urban sustainability, urban environments, democratizing local governance, water and sanitation, training and capacity building and equality and equity in municipal issues. She has organized, researched and written extensively about gender equality, diversity, and women's rights in cities. Khosla managed ICLEI – Local Governments for Sustainability's Local Agenda 21 Campaign in the 1990s. She has been a consultant for many years to the Gender and Water Alliance, UN-HABITAT and other international organizations, and was one of the founders of the Riverdale Immigrant Women's Centre and Toronto Women's City Alliance. She was the co-managing editor of *Women and Environments International Magazine* and is the author of *Gender in Local Government: A Sourcebook for Trainers*. She moderates the urban women's listserve http://groups.yahoo.com/group/urban_women.

Yianna Lambrou is senior officer in the Gender Equity and Rural Employment Division at the Food and Agriculture Organization of the United Nations (FAO) where she leads the division's work on gender and climate change. She holds a PhD in rural sociology from York University, has lectured in development studies in several universities and published extensively on natural resource management, indigenous knowledge, gender, biodiversity, bio-energy and climate change. She has consulted internationally and worked with the Canadian International Development Agency (CIDA) and with the International Development Research Centre (IDRC). She recently led the FAO research project on Gender and Climate Change in Andra Pradesh, India.

Ansa Masaud has more than seven years of experience in international development, working with civil society organizations in Pakistan and Afghanistan in setting up technical assistance programmes with the government to promote gender equality, human rights, environment, climate change and good governance. She has also participated in various networks in the Philippines and Indonesia and contributed to developing preparedness and contingency plans and systems. During the past three years she has been working with UN-HABITAT to promote an inclusive urban development agenda and her area of work includes policy and operational work in housing, land and property rights in conflict and post-conflict, humanitarian response and recovery, urban policies, governance and risk reduction, gender equality and women's rights, and sustainable urbanization. She has written numerous articles and participates in various thematic networks.

Koos Neefjes has worked on sustainable development issues in many countries over the past 25 years, in both advisory and management positions. He was policy advisor for sustainable development with Oxfam through the 1990s. He is currently policy advisor for climate change with the United Nations Development Programme (UNDP) in Vietnam, where he advises the government. He facilitates policy dialogues, performs and manages policy-relevant research, and supports donor coordination. Neefjes supports the UNDP and the wider UN in Vietnam on climate change policy work and programme development.

Biju Negi is a bilingual writer, activist and consultant, working on food security, food sovereignty, biodiversity and seed conservation, ecological agriculture and climate change, and their holistic relationships. He is closely associated with small food producers' concerns and movements – in particular with *Beej Bachao Andolan* (Save Seeds Movement), a farmers' movement in the mountain state of Uttarakhand, India. He has worked with national and international organizations.

Sibyl Nelson is a climate change adaptation and gender consultant at the FAO. Her past experience includes work with the Pew Center on Global Climate Change, Columbia University's Earth Institute, and the International Institute of Rural Reconstruction. She holds an MA in climate and society from Columbia University (focus on development) and a BA in biology and environmental studies from the University of Chicago.

Valerie Nelson is a rural social development specialist in the Livelihoods and Institutions Group, Natural Resources Institute, University of Greenwich. She has been conducting policy and action research in Latin America and Africa since the early 1990s and is currently focusing on climate change adaptation, gender and power.

Omoyemen Odigie-Emmanuel is a research and training consultant with expertise on environmental law, human rights, gender and development, and sustainable development issues. She is currently working towards a PhD at the University of Abuja, Nigeria. She has worked with the Environmental Rights Action/Friends of the Earth Nigeria as their gender officer. She is a mother and volunteers for two organizations on gender analysis, advocacy and humanitarian law documentation.

Cate Owren is programme director at WEDO, where she has focused on gender equality and climate change since 2007, leading the organization's advocacy efforts at UNFCCC negotiations and at other global fora. Prior to WEDO, Owren worked in West Africa and the Caribbean, on gender programmes in reproductive health, microfinance and fair trade. She holds a master's in socioeconomic development and bachelor's degrees in English literature and theatre.

Vijay Kumar Pandey has been with Gorakhpur Environmental Action Group (Uttar Pradesh) for more than 15 years, working with farming communities and on networking with peasants' groups, NGOs, and related organizations on sustainable agriculture, livelihood and climate change issues, improving people's perspectives. He is also the general secretary of

Laghu Seemant Krishak Morcha (Small-Marginal Farmers' Front, Uttar Pradesh), one of the largest unions of small food producers in the country.

Dana Ginn Paredes is organizing director at Asian Communities for Reproductive Justice (ACRJ). She has worked for more than 14 years as a field organizer, trainer, national programme director and organizing director. She joined ACRJ in 2003 and has overseen the development of ACRJ's current youth and worker organizing projects and climate change initiative. Paredes holds a BS in political science from the University of California at Berkeley.

Tracy Raczek is the focal point for climate change and gender at the United Nations Development Fund for Women (UNIFEM). She represents UNIFEM at international climate change negotiations and liaises with UN and civil society partners on related issues. She holds a master's in international relations with an emphasis on transboundary agreements and a bachelor's in political science, and has more than ten years in the environmental field working in forestry as well as for non-profit conservation organizations.

Ann Rojas-Cheatham has worked for the past 20 years as a community organizer, researcher, trainer and consultant. She has worked with farm workers, day labourers, young Asian women, and public health professionals on community-driven research projects to advance public health. She holds an MPH and PhD on community-based participatory research from the UC Berkeley School of Public Health in California.

Aparna Shah, formerly special projects director at ACRJ and currently executive director at Mobilize the Immigrant Vote (MIV), has worked in development, organizing and services for young people and immigrant families for more than 15 years. She joined ACRJ in 2003 as development director and transitioned to special projects director in 2007. Shah holds a master's degree in mental health from Johns Hopkins School of Public Health.

Eveline Shen is executive director at ACRJ. She has organized in communities of colour for more than 20 years. During the last ten years at ACRJ, Shen has focused her work in the Asian Pacific Islander communities and has written articles, manuals and training curricula on popular education and community organizing. She holds a master's in public health from UC Berkeley, California.

Ashbindu Singh is regional coordinator for the UNEP's Division of Early Warning and Assessment – North America, in Washington, DC. He has a strong multidisciplinary background with postgraduate degrees in physical and natural sciences and a PhD in environmental science. He has more than 100 publications in peer-reviewed scientific journals and conferences, and 35 UNEP reports on various environmental issues. The team under his direction has produced highly influential reports on various environmental issues including the UNEP's best-seller and award-winning publication *One Planet Many People: Atlas of our Changing Environment.* Singh has won numerous prestigious international awards.

Reetu Sogani is a development scholar, practitioner and consultant who, in association with small rural-based organizations, works with village and indigenous communities in the central Himalayan Uttarakhand (India) on issues of people's rights to their natural resources, local and traditional knowledge systems, sustainable agriculture and gender equity.

Jenny Svensson has seven years of academic background and professional experience in international affairs, with a main focus on development and environmental issues. In 2009, she graduated with distinction from the University of Halmstad, Sweden, with a master's degree in political science, after completing a bachelor's degree in political science and international relations in 2006. Since January 2010, Svensson has been an intern with the Delegation of the European Union (EU) to the US, working on bilateral trade and agricultural relations between the US and the EU. In 2008, she completed a five-month internship program at the UNEP Regional Office for North America, where she drafted a report on the relationship between gender and environmental changes, which served as the basis for further work in this area. From 2003–2008, Svensson served as an active member and a member of the steering committee of the UN's sub-office in Halmstad (Sweden).

Marcela Tovar-Restrepo is the current acting director of the Latin American Studies Program and the Anthropology Department at Queens College University of the City University of New York. She received her MSc in urban development planning at the University College of London and her PhD in anthropology at the New School for Social Research, New York. Tovar-Restrepo conducts research on gender, ethnicity and development in Latin America.

Katharine Vincent is interested in the human dimensions of climate change, particularly vulnerability and adaptation in southern Africa. She was a contributing author to two chapters of the Intergovernmental Panel on Climate Change (IPCC) Fourth Assessment Report (Second Working Group on impacts) and has published papers on indicators of vulnerability and adaptive capacity and institutions and governance around climate change. She now consults to public and private sector institutions on climate change, and is an associate of the ReVAMP research group at the University of the Witwatersrand, Johannesburg.

Foreword

In the nexus between gender and climate change I see the culmination of a journey of more than 25 years. During that journey we explored the many manifestations of women's roles in environmental management, and the gender aspects that interweave all aspects of sustainable development. That journey brought us to many countries, villages and communities, and it put us on the podia of national, regional and international environmental fora. We were with many female and male companions: in every corner of our globe, women and men came to understand the issues at stake, and started to learn, to share and to build awareness around gender in environment. We organized, campaigned and advocated for social and gender justice on a healthy planet. The 21st century has come with an awareness of the major challenges that climate change implies. We are bound to stop it from degrading our livelihoods and our future. We are concerned about its impacts on the growing inequality in the world. And we are convinced that a road out of this turmoil will be our combined efforts to bring gender justice and climate wisdom together in our policies, our economies and our societies. In order to make that convergence successful, my personal plea is to listen to and learn lessons from those people who live closest to the realities of today's climatic changes: women and men, boys and girls, in our world's communities, villages and towns. It is in that spirit that I would like to invite you to start this exploration with listening to the voice of one of those local leaders. I hope this book will inspire all of you, students, scientists and policy-makers, to become agents of social and climate justice.

Irene Dankelman
Malden, Netherlands
20 February 2010

Testimony

Rehana Bibi Khilji, HOPE-Pk (Humanitarian Organization for Poverty Eradication and Environment) – Pakistan (www.hope-pakistan.org)

In my life I have observed climatic changes, and I have heard my elders talk about it. About 20 years ago I could still see lots of snow in the Quetta Valley, where I was born and grew up. It was called the valley of fruit orchards; some people called it little Switzerland. I used to make statues with the snow, along with my brothers and friends. I remember when I was a child once it snowed so much that the roof of our mud house started to leak and we moved to our aunt's house for refuge.

Now it's totally changed: we hardly get any snow now. It used to rain a lot and on time, but now we get small showers two to three times a year at the most. And also some strange phenomena of fast rain in a small time scale brings disaster for people living in slum areas in their poorly made mud houses with poor sanitation and drainage facilities. I have seen our streets filled with dirty water mixed with sewage from the drains. Nowadays, more people go to the doctors for diseases such as malaria, skin infections and diarrhoea. Quetta is one of the most highly polluted cities in the world, with high levels of smoke and dust. We now have more allergies, asthma and respiratory diseases than before. Quetta, which was called the beautiful city, was developed for 50,000 people, but now it has a population of more than 1 million. The loss of livelihood opportunities in rural areas and some man-made conflicts have forced people to shift to Quetta in large numbers.

We used to have running water from the tap, but now we have to purchase water every fourth day and store it in tanks. Yet we are not sure if the water comes from a clean source. Quetta used to enjoy four seasons, but now we have longer summers and short but extreme winters, while spring has almost shrunk to a period of just one month. It is very strange for us that now we have heat shocks in a place like Quetta, which was once the destination for tourists from hot climate areas in the country.

In my own home I had a lawn, kitchen garden and sheep. But now I cannot afford to have them anymore because it is hard to purchase water for a lawn. At the same time, I know people out there who have difficulty finding water for drinking and washing.

A very painful element is that we are losing our indigenous water sources. These were once the freshwater sources for our communities and we used them for irrigating our land. During my field visits I have seen that underground water levels have gone down. This is due to multiple factors, such as deforestation, lack of rain and snow, water extraction through electric pumps and lack of recharge mechanisms. Now women in the village have to travel a long way to fetch water. The small birds that had nests in the mud walls and used to lay eggs are hardly seen anymore.

What I really miss is that I do not see the indigenous women selling herbal medicines that they used to collect from rural areas in Baluchistan. We used to apply them for treatment of health problems, such as coughs, colds, flu, burns, and hair and skin care. Now due to drought and loss of rangelands and green cover, we miss not only these products, but also the local skills that were associated with them.

Women used to make cheese and yogurt from milk of the livestock – now it's almost all gone. They also used to make carpets and rugs. All such products have to be purchased from the market now, and at the same time they have lost a basic source of income and empowerment associated with their income. Not only this, but also their source of nutrition has been affected: in poor families pregnant and lactating women are now mostly undernourished. All this I have seen myself while working with these communities.

I have seen families who were prosperous 15 years ago now forced by drought to sell their orchard lands for other purposes at small sums. Even the quality of fruits such as apples, pears and plums is not the same as before. In the south of Baluchistan now I see more floods happening each year, and these become a regular phenomenon, with severe impacts on the poorest households and particularly women. We women suffer a lot because we are the primary caregivers and handlers of the environment and ecosystems at the grassroots level.

I can see that it is not only the lack of proper planning at the local level that has caused these problems – there are major contributors at global level whose unjust drive for more unfair policies have disturbed the balance of our larger ecosystem. So the ones responsible should take informed and just actions to mitigate climate change and to help us adapt better to its impacts.

Since Rehana Khilji contributed her testimony to our book, Pakistan and the region have been struck by the worst flooding ever. This book is dedicated to all local women and men, girls and boys, who are suffering so much from climate-induced disasters and calls on everybody to support humanitarian action, climate change adaptation, prevention and reconstruction.

List of Acronyms and Abbreviations

ACCRN	Asian Cities Climate Change Resilience Network
ACRJ	Asian Communities for Reproductive Justice (USA)
ANGRAU	Acharya NG Raya Agricultural University (India)
AWID	Association of Women in Development
BNRCC	Building Nigeria's Response to Climate Change
BOTEC	Botswana Technology Centre
BPC	Botswana Power Corporation
BRT	Bus Rapid Transit
CAN	Climate Action Network
CBDRM	community-based disaster risk management
CBO	community based organization
CDM	Clean Development Mechanism
CEDAW	Convention on the Elimination of all forms of Discrimination Against Women
CIDA	Canadian International Development Agency
CIF	Climate Investment Fund
COHRE	Centre for Housing Rights and Evictions
COP	Conference of Parties
CRIAW	Canadian Research Institute for the Advancement of Women
CSD	Commission on Sustainable Development (UN)
CSE	Centre for Science and Environment
CSIR	Council for Scientific and Industrial Research (South Africa)
CSW	Commission on the Status of Women (UN)
CWC	Coastal Women for Change (US)
CWGL	Center for Women's Global Leadership
DECRIPS	Declaration on the Rights of Indigenous Peoples (UN)
DPSIR	drivers–pressures–state–impact–response
DRM	disaster risk management
ECOSOC	Economic and Social Council (UN)
EECCA	Eastern Europe, Caucasus and Central Asia
EPA	Environmental Protection Agency
EU	European Union
FAO	Food and Agriculture Organization of the United Nations

GBM	Green Belt Movement
GDN	Gender and Disaster Network
GDP	gross domestic product
GEAR	Gender Equality Architecture Reform
GEF	Global Environment Facility
GGCA	Global Gender and Climate Alliance
GHG	greenhouse gas
GLTN	Global Land Tools Network
GRHS	Global Report on Human Settlements
GWP	global warming potential
HFA	Hyogo Framework for Action
ICCE	Indigenous Climate Change Ethnographies
ICIMOD	International Centre for Integrated Mountain Development
IDRC	International Development Research Centre
IFAD	International Fund for Agricultural Development
ILO	International Labour Organization
IMF	International Monetary Fund
IPCC	Intergovernmental Panel on Climate Change
ISDR	International Strategy for Disaster Reduction
IUCN	International Union for Conservation of Nature
JI	Joint Implementation
LCA	life cycle assessment
MDG	Millennium Development Goal
MOV	Mobilize the Immigrant Vote
NAPA	National Adaptation Plan for Action
NBSAP	National Biodiversity Strategy and Action Plan
NEST	Nigeria Environmental Study/Action Team
NFWI	National Federation of Women's Institutes
NGO	non-governmental organization
NTFP	non-timber forest product
NWMLE	Network of Women Ministers and Leaders for the Environment
OECD	Organisation for Economic Co-operation and Development
PFCs	perfluoro compounds
POLISH	Participatory Research on Organizing Leadership for Safety and Health
R&D	research and development
REDD	Reducing Emissions from Deforestation and Forest Degradation
SIDA	Swedish International Development Agency
SRHR	sexual and reproductive health and rights
SRI	System of Rice Intensification

TCO_2e	Tonnes of CO_2 equivalent
UHN	United Houma Nations (US)
UN	United Nations
UNAPA	Unión de Asociaciones Productivas del Altiplano (Bolivia)
UNCBD	United Nations Convention on Biological Diversity
UNCCD	United Nations Convention to Combat Desertification
UNCED	United Nations Conference on Environment and Development
UNDP	United Nations Development Programme
UNEP	United Nations Environment Programme
UNESCO	United Nations Educational, Scientific and Cultural Organization
UNFCCC	United Nations Framework Convention on Climate Change
UNFPA	United Nations Population Fund
UN-HABITAT	United Nations Human Settlements Programme
UNIFEM	United Nations Development Fund for Women
UNSD	United Nations Statistics Division
UN-UDHR	United Nations Universal Declaration on Human Rights
VOC	volatile organic compounds
WAVE	Women As the Voice for the Environment
WECF	Women in Europe for a Common Future
WEDO	Women's Environment & Development Organization
WEN	Women's Environmental Network
WHJI	Women's Health and Justice Initiative (US)
WHO	World Health Organization
WMO	World Meteorological Organization
WOCAN	Women Organizing for Change in Agriculture and Natural Resources Management
WRI	World Resources Institute

Introduction: Exploring Gender, Environment and Climate Change

Irene Dankelman

This chapter is an introduction to the challenging nexus of gender and climate change. It will describe the main developments in practice and thinking around these themes, with an emphasis on developments within the past 25 years. It will also explain the main concepts and key definitions, and give some guidance on the structure of the book.

Introduction

Gender and Climate Change: An Introduction offers an analysis of the interface between the changing physical environment and human society, with a particular focus on gender aspects of climate change. More than 20 years ago, in 1988, the first textbook on women and environment was published (Dankelman and Davidson, 1988). Since that time, almost every aspect of the women–environment nexus has undergone important changes. One of the most pressing environmental challenges is currently climate change, exacerbating all the environmental issues of the late 1980s: water, energy, land use and biodiversity conservation. At the same time, human factors have undergone major changes and the relationships between women and men worldwide manifest themselves as diverse and dynamic gender characteristics. Changes in the world's climatic conditions, the way these are formed, and how people are affected, cope with and adapt to these, have important and challenging gender dimensions. This book aims to introduce these dimensions and interactions, and intends to build a broad understanding of gender aspects of climatic change. First it addresses students and professionals active in areas such as environment, climate change, social justice and sustainable development. Secondly it intends to inform a broader, interested public.

As global environmental change, and in particular climate change, is a major challenge for our societies today and impacts all our efforts to build a more just and sustainable society, the book underlines the need for a gender-specific approach in climate change policies, actions, study and research. Based on a wide range of experiences on the ground, it will argue that men and women alike are important as agents of change and as distinct counter-forces to cope with and adapt to processes of global and local environmental change, and to mitigate these. These societal roles and functions need recognition as well as mainstreaming of a gender perspective in climate change policies, mechanisms and actions.

Acknowledgements

This book is developed with the inputs of a wide range of experts from all over the world. Particular thanks go to those who cooperated in and contributed to the diverse chapters in this publication: Lorena Aguilar, Eleanor Blomstrom, Willy Jansen, Gerd Johnsson-Latham, Prabha Khosla, Ansa Masaud, Cate Owren and Tracy Raczek. They have brought their profound knowledge and expertise in specific areas of gender equality and climate change to the forefront. The editor is also particularly thankful to the many professionals who, through their testimonies and case studies, have given visibility to the realities on the ground, where women and men live their lives, experience climate change impacts and take action: Emma Archer, Bernhard Barth, Sabine Bock, Ruth Bond, Bharati Chaturvedi, Emily Cleevely, Thais Corral, Tracy Cull, Gero Fedtke, Sascha Gabizon, Shana Griffin, Rachel Harris, Janet Kabeberi-Macharia, Rehana Bibi Khilji, Yianna Lambrou, Koos Neefjes, Biju Negi, Sibyl Nelson, Valerie Nelson, Omoyemen Odigie-Emmanuel, Vijay Kumar Pandey, Dana Ginn Parades, Ann Rojas-Cheatham, Aparna Shah, Eveline Shen, Ashbindu Singh, Reetu Sogani, Jenny Svensson, Marcela Tovar-Restrepo and Katherine Vincent. It is through the leadership of all these contributors that we can be inspired, and it is for this reason that their short biographies are also included in the book.

Exploring the issues:
A bird's-eye view through history

The interface between ecological and social processes is not only maintained through natural processes, products and impacts of the environment on human beings, but also by the endeavours of humans themselves. Since the existence of humankind, people have interacted with the physical environment in direct or indirect ways. In hunter-gatherer societies this interaction was based on the principle of the harvesting of natural products. These communities must have developed a deep understanding of the characteristics of ecosystems and their resources. In the early settled agricultural societies, humans started to cultivate the land and other resources in order to gain enough food and other produce from the reshaped environment, and started building up stocks, originally for their own use, and later for local (and even international) markets. Experiences, beliefs and visions resulted in specific management practices of the physical environment.

The notion that human interventions and overuse of natural resources – often forests – could result in a disturbance of natural cycles and in human misery already existed in Greek and Roman times. Plato (430–373 BC) warned in his *Kritias* that the deforestation of the mountains around Athens would result in loss of the water-storage capacity of the environment and would finally end with erosion and flooding. This would result in the further degradation of the once fertile agricultural lands of Attica, and the drying up of water sources. The Roman historians Strabo (64/63 BC–23 AD) and Plinius (23–79 AD) also questioned the unlimited deforestation for mining, shipbuilding and construction work, and its consequences for water retention and land management. This awareness of limited goods finally resulted in the notion of sustainable development, as developed by the German mining engineer Von Carlowitz in 1713 in his *Sylvicultura Oeconomica* on sustainable forest use (van Zon, 2002).

This perspective did not only emerge in the Western world. A well-known early example of people's awareness of the limitations of natural capacities and the need to safeguard these for human well-being is that of the Bishnoi community in Rajasthan, India. The Bishnoi are known as environmental conservationists and managers in the Thar desert. In 1730 the Maharaja Abhay Singh of Jodhpur, after inviting local men to a party, ordered his soldiers to fell trees in a Bishnoi village and bring back wood for the construction of a new palace. But when Amrita Devi from that village came to know about those plans she organized the villagers and they went to protect the trees with their own bodies. In the Khejadli Massacre, Amrita Devi, along with 362 other local villagers, died while protecting the khejri trees (*Prosopis cineraria*). As soon as the Maharaja learned about the massacre, he ordered the stoppage of the tree-felling (Mukhopadhyay, 2008) (see also Chapter 9).

Overuse of natural resources and also pollution are as old as human history. Increasing human populations, particularly in early urban settlements, opened the door to more pollution and disease. During the Middle Ages, diseases such as cholera and typhoid fever broke out all across Europe and were directly related to unsanitary conditions and human and animal waste. By the mid-1800s, people began to understand that unsanitary conditions and water contamination adversely affected human health. This 'new' awareness prompted major cities to take measures to control waste and garbage (Goubert, 1989).

With the advancing industrial development in the 19th century, human interventions in the physical environment intensified, with increasing demands for wood, coal, minerals, crops and other natural products. In the Western world, signs of overuse in agriculture, forestry and mining, as well as pollution of water and air, became more and more visible. Some scholars warned that the natural environment was not infinite, and that use of natural resources should be guided by some form of wise use, or sustainability (van Zon, 2002). However, it was not until the late 19th century – with the establishment of nature conservation organizations – and the mid-20th century – with the release of publications such as the book *Silent Spring* by Rachel Carson (1962), the report *Only One Earth: The Care and Maintenance of a Small Planet* by Barbara Ward and René Dubos (1972) and *The Limits to Growth: A Report to the Club of Rome* (Meadows et al, 1972) that the awareness of the need for environmental management and nature conservation became a more common good and was perceived as a global responsibility. Since that time, the study of the interactions between the human and physical environments started to develop, and became an object of environmental sciences.

People's interaction with the natural environment is as old as humankind itself, but it has changed over time. Women and men have played diverse roles in this interface. The gender-specific roles, rights and responsibilities of people in their physical environment were first highlighted by scholars such as Ester Boserup (1970, 1989) – on agriculture – and organizations such as the Food and Agriculture Organization of the United Nations (FAO), with regard to agriculture and forestry, and the International Union for Conservation of Nature (IUCN) on biodiversity conservation.

In her essay 'Is Female to Male as Nature to Culture?' the feminist anthropologist Sherry Ortner (1974) analyses the secondary status of women in society as a universal phenomenon, based on the pan-cultural assumption that women are closer to nature than men, with men seen as occupying the higher ground of culture. Women occupy the intermediary space between nature and culture. While rejecting biological determinism, Ortner bases her assumptions on women's physiology and social roles, as well as their psyche. Critics argue, however, that Ortner's arguments further strengthen the dichotomy between men and women, and that cultural diversity does not allow for any universal pronunciations. Later in her essay 'Gender Hegemonies' (1990) Ortner acknowledges that her critics were more right than she initially admitted. She developed a more complex picture of dominance, seeing societies in which women have strong roles. However, Ortner's focus on women's subordination in society does not question the societal attitude to nature itself (as being of a lower value than culture), as did Carolyn Merchant.

In her book *The Death of Nature: Women, Ecology and the Scientific Revolution* (1980) science historian Carolyn Merchant argues that there is a major parallel between the degradation of nature and the oppression of women. One of the major causes, she argues, lies in men's changing valuation of nature during the Enlightenment, when they began seeing nature as something to be used, explored and exploited. Similarly women were perceived as having inferior and serving positions in communities and households. This notion of the parallel between women's and nature's positions in society is further explored by later eco-feminist thinkers. Critics warn, however, of over-simplification of such relationships, and against the dangers of comparing the female body with mother nature (Braidotti et al, 1993). However, the realities in which many women and men live show more and more clearly the specific differential roles, rights and responsibilities of men and women in managing and maintaining the physical environment. Based on such observations since the mid-1980s the specific nexus between gender, environment and sustainable development has been explored by scholars, activists and development workers (CSE, 1985; Rocheleau, 1985; van Wijk-Sijbesma, 1985; Agarwal, 1986; Cecelski, 1986; Dankelman and Davidson, 1988; Shiva, 1988; ICIMOD, 1988; Monimart, 1989). Many of these studies show that women's positions and roles have been seriously neglected, not only in the practice of the environmental conservation and sustainable management of resources, but also in the more scientific foundations of such activities in environmental science and studies. At policy level, the need to mainstream gender in the environmental sector and in sustainable development efforts has been recognized during the past 15–20 years, although often reluctantly. This recognition is still not internalized in many institutions and needs intense external advocacy work in order to result in gender-sensitive policies and practices of dominant organizations and institutions. This is especially the case with regard to gender and climate change, as will be described in the following chapters.

Many of the existing studies and research on gender and environment focus on specific sectors, elements and cases, such as gender roles in food production, in water management and in energy use at global, national or local level. Deep analyses and analytical frameworks to understand the nexus gender–climate change–sustainable development at meta-level are still limited. Similarly, a genuine debate on the relevance of that nexus for climate change science and practice is almost absent. This book shows that in arenas that deal with climate change and sustainable development, there is still too limited reference made to the gender dimensions. On the other hand, many examples show that if women participate in climate change action, planning and decision-making, they contribute valuable knowledge, experiences and perspectives, and they can take on important leadership roles. Some of these cases are analysed in this book. It shows that a far more fundamental understanding, policy development and practice on gender–climate change and sustainable development is needed.

In the meantime, as for example the Millennium Ecosystem Assessment (2005) and the Fourth Assessment Reports of the International Panel on Climate Change (IPCC, 2007) show, the environment changes at an unprecedented rate due to human interventions. These changes have major impacts on both human livelihoods and human lives. There is an urgent need to understand these changes and their impacts from a gender perspective, and to draw lessons from that understanding. This book hopes to contribute to that notion by offering students, scholars and policy-makers a basic body of knowledge on gender dimensions of climate change.

Concepts and definitions:
From climate and climate change

Main concepts that are used in this book are defined here. The book talks about the physical environment or the eco-sphere: the air, water, land, flora and fauna, in their respective ecosystems. In order to distinguish the non-human and the human spheres of life, the human society is described as the socio-sphere. Focus of the book will be formed by the differential processes in the interface between the socio-sphere and eco-sphere, with an emphasis on climatic changes and gender issues, and there it enters the domain of human ecology.

Weather conditions have a major impact on plant growth, animal life and human well-being. 'Climate' in a narrow sense is usually defined as the 'average' weather. More rigorously, it stands for the statistical description in terms of mean and variability of relevant variables over a period of time, ranging from months to thousands or millions of years, but with as classical period of time 30 years (World Meteorological Organization). These variables are often temperature, precipitation and wind. In a wider sense, climate is the state of the climate system. According to the IPCC, 'climate change' refers to a change in the state of the climate that can be identified by changes in the mean climate and/or the variability of its properties, and that persists for an extended period, typically decades or longer (IPCC, 2007). Climate change may be due to natural internal processes and fluctuations, or to persistent anthropogenic changes, coming from the socio-sphere with manifestations in the composition of the atmosphere or in land use. The United Nations Framework Convention on Climate Change (UNFCCC), in its Article 1, takes a specific focus by defining climate change as: 'A change of climate which is attributed directly or indirectly to human activity that alters the composition of the global atmosphere and which is in addition to natural climatic variability observed over comparable time periods.' In this book, the IPCC definition will be followed, particularly in the chapters on climate change impacts and adaptations to climate change. But in sections on mitigation and policy development, the more focused approach of the UNFCCC will be kept to, underlining possibilities to mitigate human-induced and enhanced climatic changes.

Changes in atmospheric concentrations of greenhouse gases (GHGs) and aerosols, land cover and solar radiation alter the energy balance of the climate system (IPCC, 2007). GHGs are gases in the Earth's atmosphere that absorb and re-emit infrared radiation. These gases occur through both natural and human-influenced processes. Primary GHGs include carbon dioxide (CO_2), nitrous oxide (N_2O), methane (CH_4), ozone and CFCs. Climatic changes manifest themselves into extreme weather events – such as storms and cyclones, cold or heatwaves, heavy rainfall or droughts – and in slower processes of climatic changes, such as changes in precipitation, average temperatures or changes in wind patterns. Climate change results in melting of snow and ice, sea levels rising, drying up of rivers and aquifers, wildfires, bleaching of coral reefs and an increase in pathogens and illnesses. These have major effects on ecosystems, agriculture, forestry and fisheries, on water resources and industry, and on human health, settlements and society.

In order to reduce, delay or avoid the impacts of climate change adaptation and mitigation, measures are taken by governments, communities and other stakeholders. Adaptation in this context is seen as the adjustment in natural and human systems in response to actual or expected climatic stimuli or effects, which moderates harm or exploits beneficial opportunities. Various types of adaptation are distinguished, including anticipatory, autonomous and planned adaptation. Mitigation, on the other hand, is a human intervention to reduce anthropogenic forcing of the climate system. It includes strategies to reduce GHG sources and emissions and to enhance GHG sinks (IPCC WG2, 2007).

Box 1.1 *Main findings of the Fourth Assessment Report of the IPCC (2007)*

Observed changes in climate:

- Warming of the climate system is unequivocal, as is now evident from observations of increases in global average air and ocean temperatures, widespread melting of snow and ice and rising global average sea level.

- Observational evidence from all continents and most oceans shows that many natural systems are being affected by regional climate changes, particularly temperature increases.

- Other effects of regional climate change on natural and human environments are emerging, although many are difficult to discern due to adaptation and non-climatic drivers.

Causes of change:

- Global GHG emissions due to human activities have grown since pre-industrial times, with an increase of 70 per cent between 1970 and 2004.

- Global atmospheric concentrations of carbon dioxide, methane and nitrous oxide have increased markedly as a result of human activities since 1750 and now far exceed pre-industrial values determined from ice cores spanning many thousands of years.

- Most of the observed increase in global average temperatures since the mid-20th century is *very likely* due to the observed increase in anthropogenic GHG concentrations.

- It is *likely* that there has been significant anthropogenic warming over the past 50 years averaged over each continent (except Antarctica).

- Anthropogenic warming over the last three decades has *likely* had an apparent influence at the global scale on observed changes in many physical and biological systems.

Projected climate change and its impacts:

- There is *high agreement and much evidence* that with current climate change mitigation policies and related sustainable development practices, global GHG emissions will continue to grow over the next few decades.

- Continued GHG emissions at or above current rates would cause further warming and induce many changes in the global climate system during the 21st century that would *very likely* be larger than those observed during the 20th century.

- There is now higher confidence than in the Third Assessment Report in projected patterns of warming and other regional-scale features, including changes in wind patterns, precipitation and some aspects of extremes and sea ice.

- Anthropogenic warming and sea level rise would continue for centuries due to the time scales associated with climate processes and feedbacks, even if GHG concentrations were to be stabilized.

- Anthropogenic warming could lead to some impacts that are abrupt or irreversible, depending upon the rate and magnitude of the climate change.

Adaptation and mitigation options:

- Adaptation can reduce vulnerability, both in the short and the long term.

- A wide array of adaptation options is available, but more extensive adaptation than is currently occurring is required to reduce vulnerability to climate change.

- Adaptive capacity is intimately connected to social and economic development, but is unevenly distributed across and within societies.

- Both bottom-up and top-down studies indicate that there is *high agreement* and *much evidence* of substantial economic potential for the mitigation of global GHG emissions over the coming decades that could offset the projected growth of global emissions or reduce emissions below current levels.

- There is *high agreement* and *much evidence* that a wide variety of policies and instruments are available to governments to create the incentives for mitigation action. Their applicability depends on national circumstances and understanding of their interactions, but experience from implementation in various countries and sectors shows that there are advantages and disadvantages for any given instrument.

- There is also *high agreement* and *medium evidence* that changes in lifestyle and behaviour patterns can contribute to climate change mitigation across all sectors. Management practices can also have a positive role.

Longer-term perspective:

- Responding to climate change involves an iterative risk management process that includes both mitigation and adaptation, taking into account actual and avoided climate change damages, co-benefits, sustainability, equity and attitudes to risk.

- There is *high confidence* that neither adaptation nor mitigation alone can avoid all climate change impacts. However, they can complement each other and together can significantly reduce the risks of climate change.

- Many impacts can be reduced, delayed or avoided by mitigation. Mitigation efforts and investments over the next two to three decades will have large impacts on opportunities to achieve lower stabilization levels. Delayed emission reductions significantly constrain the opportunities to achieve lower stabilization levels and increase the risk of more severe climate change impacts.

- Sustainability can reduce vulnerability to climate change, and climate change could impede nations' abilities to achieve sustainable development pathways.

Source: IPCC (2007)

Concepts and definitions: About gender, women and gender equality

'Gender' is seen as the manifestation of the dynamic and context-specific relationships between women and men. This book identifies 'gender' as 'the socially acquired notions of masculinity and femininity by which women and men are identified' and 'gender relations' as 'the socially constructed form of relations between women and men'(Momsen, 2004, p2). Often gender is understood as the cultural difference of women and men, based on the biological division between male and female. However, there are objections to this definition as it simply divides human society into two realms, and by its dichotomy excludes the patterns of difference among women and among men. Moving from a focus on difference the focus on relations evolved (Connell, 2002). Defined that way, gender becomes a dynamic ideological and cultural construct that is also reproduced within the realm of material practices, giving meaning to sex differences. At the same time gender is shaped by the cultural notions of masculinity and femininity. Davids and Van Driel (2002) distinguish different dimensions of gender – that constantly interact with each other: the symbolic (cultural texts, representations, stereotypes), the structural, institutional dimension (in which the structural differences amongst women become visible, such as labour relationships), and the individual subject (the way individuals express their identity). Gender affects the distribution of resources, wealth, work, decision-making and political power, and the enjoyment of rights and entitlements in all spheres (UN, 1999).

Gender relations are contextually specific and often change in response to altering circumstances (Moser, 1993, p230). Despite variations across cultures, regions and over time, gender relations throughout the world entail asymmetry of power between women and men as a pervasive trait. Indeed, gender inequality is one of the most pervasive of all inequalities (UNDP, 2005). Thus, gender is a social stratifier, and in this sense it is similar to other stratifiers such as race, class, ethnicity, religion, place and age, that in themselves all affect gender roles and meanings. This means that the differences among women also need specific consideration. Although this book tries to do justice to these differentials in the course of its text, generalizations about women will be made. Or as Zillah Eisenstein concluded: 'However differentiated gender may be, gender oppression exists'(1994, p8). And so in many societies women as a group can still be seen as having subordinate positions and neglected capacities.

The book will look into 'gender' to focus on aspects of (in)equality between men and women, and their respective roles, positions and needs. But it will also focus on women as such, in order to give visibility to their respective roles, positions and needs. In this approach we follow scholars such as Harcourt and Escobar (2005, p2), who begin their publication *Women and the Politics of Place* with the remark that they are deliberately focusing on women rather than on feminist analysis or gender relations. They see the political importance of looking at women, beginning with how women themselves experience it.

> *Too often the differences for women and men become smoothed away in progressive analytical frameworks. Knowledge about women continues to be hardest to come by, and although many of us work in feminist theory, we try in this book not to assume that readers share that knowledge, but instead bring it in when it helps explain the story we are telling.* (Harcourt and Escobar, 2005, p2)

A similar argumentation to bring the focus back from gender relationships to women as such, is stressed by Devaki Jain in her book *Women, Development and the United Nations* (2005). Jain describes how, over the years, language to speak about women has changed several times. She returns to the word 'women, or the women word', as an attempt to envelop the political identity of 'women'. 'Now in 2005 an increasingly accepted view is that to reclaim political identity, to affirm women's collective political will, the word "women" has returned as preferred currency' (Jain, 2005, p5). In order to give visibility to women's roles and positions in the socio-ecological spheres of our societies we will refer to 'women' as (groups of) persons and to 'gender' as a construct. Thereby the structural relationships, including labour inputs, physical characteristics, rights and access to and control over resources, level of decision-making power, as well as cultural aspects and identities, are important gender-determinants to understand the socio-ecological interactions.

In order to ensure gender equality and to overcome problems of marginalization, invisibility and under-representation, gender concerns and women's issues have been integrated into mainstream policies, programmes and projects, and in institutional structures and procedures through gender mainstreaming (Vainio-Matilla, 2001). The definition of the Group of Specialists on Gender Mainstreaming at the Council of Europe has been widely adopted because it accentuates gender equality as an objective, and not women as a target group, and because it emphasizes that gender mainstreaming is a strategy. This definition says that: 'Gender mainstreaming is the (re)organization, improvement, development and evaluation of policy processes, so that a gender equality perspective is incorporated in all policies at all levels and at all stages, by the actors normally involved in policy-making' (Council of Europe, 1998, p15). Widely accepted is the definition developed by the Economic and Social Council (ECOSOC) in its *Agreed Conclusions* 1997/2:

> *Mainstreaming a gender perspective is the process of assessing the implications for women and men of any planned action, including legislation, policies or programmes, in all areas and at all levels. It is a strategy for making women's as well as men's concerns and experiences an integral dimension of the design, implementation, monitoring and evaluation of policies and programmes in all political, economic and societal spheres so that women and men benefit equally and inequality is not perpetuated. The ultimate goal is to achieve gender equality.*

Gender mainstreaming can be best understood as a continuous process of infusing both the institutional culture and the programmatic and analytical efforts of agencies with gendered perspectives. This implies fundamental changes, not only in the organization of work, but also in the work itself and its quality. 'Gender mainstreaming means taking gender seriously –

and taking it into account in all aspects of the workplace and the work products of an institution'(Seager and Hartmann, 2005, p3). Both authors distinguish several conditions for success of gender mainstreaming: (a) an institutional culture that is open to gender perspectives, and willing to undertake self-assessment; (b) political commitment at the highest level; (c) gender mainstreaming is understood as a continuous, fluid and evolving responsibility; (e) careful use of available gender-differentiated data, indicators and analysis, (f) deployment of adequate resources (human and financial). The present situation in many organizations and institutions, however, shows that gender mainstreaming is lacking behind. Seager and Hartmann relate this to: (a) a hostile and indifferent institutional culture; (b) the 'ghettoziation' of gender; (c) the framing of gender mainstreaming as a single and finite target; (d) the inadequacies of indicators, data, and analyses that reveal gendered dimensions of issues or that support gender-disaggregated data. Rao and Kelleher (2005) in their article 'Is there life after gender mainstreaming?' point to the fact that, while women have made many societal gains in the last decade, the successful promotion of women's empowerment and gender equality are not institutionalized yet in the day-to-day routines of states, agencies and other institutions. 'More important are the myriad, insidious ways in which the mainstream resists women's perspectives and women's rights' (Rao and Kelleher, 2005, p58). There is a fundamental difficulty in transforming the paradigm of patriarchy and the gender bias of institutions into a culture, structure and operational way of work that enhances gender equality and women's empowerment. With Levy (1996), Rao and Kelleher argue that effective gender mainstreaming needs transformational institutional change. This is also true for epistemologies and dominant knowledge systems; paradigm shifts are needed to make these gender-inclusive.

Climate change and gender

The book *Women and Environment in the Third World: Alliance for the Future* (Dankelman and Davidson, 1988) focused on women's roles and positions in land use, water management, forest use, energy provision and use, urban development and conservation. It is women's diverse reproductive and productive roles in these areas in the household and at community level that are impacted adversely by climatic changes. Floods, droughts, cold and heatwaves, cyclones and hurricanes, higher average temperatures, wildfires and sea level rise: these all have major impacts on people's lives and livelihoods, and particularly on those for which women are responsible. In most cases, resources get scarcer, production goes down, prices go up and conflicts regarding resources increase – phenomena that we have already seen over the past few years. This all weighs heavily on women's shoulders and human security is at stake.

> *I have seven children... The floods collapsed our three rooms and washed away our crops: maize and late millet. As a result we harvested nothing. Hunger stared us in the face.* (Atibzel Abaande, Bawku West District, Ghana in Mensah-Kutin, 2008, p31)

Like many disasters, climate change threatens to increase existing inequalities, and – as the United Nations Development Programme (UNDP) mentioned in its Human Development Reports of 2005 and 2007 – gender inequality is one of the most pervasive of these. Important lessons can be learned from what we already know from past (natural) disasters. The socially constructed gender-specific vulnerability of women leads to the relatively higher female disaster mortality rates compared to those of men (Neumayer and Plümper, 2007).

Being in a disadvantaged position, women face specific difficulties in dealing with disasters. For example, lower levels of education and training can reduce their ability to access necessary information before, during and after disasters. And when poor people – many of whom are women – lose their livelihoods, they slip deeper into poverty. As climate change is not gender neutral, gender-specific human security impacts can already be observed in many localities where climate change hits.

Women, like men, have developed several coping strategies. A warning here: 'coping' is not always sustainable or the best way to deal with the complex problems, but merely a survival strategy for the short term. As there often is no choice, measures such as adaptations in diets, longer working days, shifting to other fuels, employment under unsafe conditions or even forced migration are all coping strategies that women delve into.

Too often, women are primarily perceived as victims of climate change, with men as actors. However, there are cases in which we have seen men as victims and often women can be positive agents of change. Because of their work on the land and dependency on natural resources, many women have acquired knowledge of local circumstances and changes. Too often planners and decision-makers do not consider these contributions adequately yet (WEDO, 2007).

As underlined by Enarson (2000) and O'Brien (2007) natural disasters, and therefore also those induced by climatic changes, can provide women with an unique challenge to change their (gendered) status in society as well. There are cases from which it becomes clear that women have been very effective in mobilizing communities to respond to disasters and in disaster preparedness and mitigation. Many women worldwide have ideas on how to mitigate and adapt to climate change, and they organize themselves in order to have their voices heard. Apart from local and regional movements of women dealing with environment and climate change, in recent years there has also been an increase of organizations active in the interface between gender equality and climate justice at global level.

Gender mainstreaming in climate change is slowly taking shape in international climate change arenas. It is not only a question of having more women with different backgrounds participating and having a say in climate change negotiations and decisions at national and global levels, there is also a need for climate change policies and practices themselves to be sustainable and just. Or as Bella Abzug, former US congresswoman and founder of the Women's Environment & Development Organization (WEDO), once said: 'Women do not want to be mainstreamed into a polluted stream: they want the stream to be clear and healthy.'

Outline of the book

After this introduction to the issues, in the following chapter an analytical framework on the multifaceted nexus gender, environment and climate change will be presented. It gives an extensive description of how gender aspects relate to the physical environment and to the use and management of natural resources. It also focuses on the impacts of environmental – including climatic – changes on gender equality. In its final part, a strategic approach to deal with the problems identified will be introduced. Chapter 3 looks into gender-specific impacts of climatic changes, from a vulnerability and capability perspective. It does so by emphasizing the human security aspect in the context of climate change. A case study from the north of India illustrates these challenges. Chapter 4 describes the relationship between gender equality and climate change in urban areas. An important argument is that the majority of the world's population lives in urban environments and faces major challenges with regard to climate change adaptation and mitigation. A case study describes the important roles in GHG mitigation of waste recyclers.

In the next section of the book – in Chapter 5 – a wide variety of realities on the ground are described. Not only are assessments of gender specific impacts of climatic change in almost all regions in the world presented, but the case studies also describes women's and men's agency in climate change adaptation and mitigation.

The final section of this publication, on strategies and actions, starts with Chapter 6, which looks into gender aspects of climate change adaptation and mitigation. It specifically examines the actual and potential impacts and contributions of climate change mechanisms and funding for gender equality. Chapter 7 deals with policy frameworks, including soft and hard law, that can contribute to making climate change policies and actions more gender responsive and sensitive. Chapter 8 looks into the gender aspects of consumption patterns and the impacts for climate change policies and action. Chapter 9 gives an overview of the mobilization and organization of women worldwide for a healthy environment and safe climate. Finally, Chapter 10 reflects on lessons learned and elaborates on critical elements – such as access to information and sex-differentiated data – and strategic approaches for strengthening gender equality in the context of climate change.

References

Agarwal, B. (1986) *Cold Hearts and Barren Slopes: The Woodfuel Crisis in the Third World*, Zed Books, London

Boserup, E. (1970, 1989) *Women's Role in Economic Development*, Earthscan, London

Braidotti, R., Wieringa, S., Hausler, S. and Charkiewics-Pluta, E. (1993) *Women, Environment and Sustainable Development: Towards a Theoretical Synthesis*, Zed Books, London

Carson, R. (1962) *Silent Spring*, Fawcett Publications, Greenwich, CN

Cecelski, E. (1986) 'Energy and rural women's work, Conference Proceedings, ILO, 21–25 October 1986', International Labour Organization, Geneva

Connell, R. W. (2002) *Gender*, Polity Press, Cambridge

Council of Europe (1998) *Gender Mainstreaming: Conceptual Framework, Methodology and Presentation of Good Practice*, Council of Europe, Strasbourg, p15

CSE (Centre for Science and Environment) (1985) 'Women and natural resources', in CSE (ed) *State of the Indian Environment: Second Citizens Report 1984–1985*, CSE, Delhi

Dankelman, I. and Davidson, J. (1988) *Women and Environment in the Third World: Alliance for the Future*, Earthscan, London

Davids, T. and Van Driel, F. (2002) 'Van vrouwen en ontwikkeling naar gender en globalisering', in B. Arts, P. Hebinck and T. Van Naerssen (eds) *Voorheen de Derde Wereld – Ontwikkeling anders Gezien*, Mets & Schilt, Amsterdam

Eisenstein, Z. (1994) *The Color of Gender: Reimaging Democracy*, University of California Press, Berkeley

ELC (Environment Liaison Centre) (1985) 'Women and the environmental crisis', report of the Workshops on Women, Environment and Development, Nairobi, 10–20 July 1985, ELC, Nairobi

Enarson, E. (2000) 'Gender and natural disasters, IPCRR Working Paper no1', International Labour Organization, Geneva

Goubert, J. P. (1989) *The Conquest for Water*, Princeton University Press, Princeton, NJ

Harcourt, W. and Escobar, A. (2005) *Women and the Politics of Place*, Kumarian Press, Bloomfield

ICIMOD (International Centre for Integrated Mountain Development) (1988) 'Women in Mountain Development, Report of the International Workshop on Women, Development, and Mountain Resources: Approaches to Internalising Gender Perspective, Kathmandu, 21–24 November 1988', ICIMOD, Kathmandu

IPCC (Intergovernmental Panel on Climate Change) (2007) *Fourth Assessment Report: Climate Change 2007. Synthesis Report. Summary for Policy Makers*, IPCC, Geneva

IPCC WG2 (Working Group II) (2007) *Climate Change 2007: Impacts, Adaptation and Vulnerability: Contribution of Working Group II to the Fourth Assessment Report of the Intergovernmental Panel on Climate Change*, M. L. Parry, O. F. Canziani, J. P. Palutikof, P. J. van der Linden and C. E. Hanson (eds), Cambridge University Press, Cambridge

Jain, D. (2005) *Women, Development and the United Nations*, Indiana Press, Bloomington, Indianapolis

Levy, C. (1996) 'The process of institutionalising gender in policy and planning – the "web" of institutionalisation', in *Working Papers*, University College London, Development Planning Unit, London

MEA (Millennium Ecosystem Assessment) (2005) *Ecosystems and Human Well-Being: Synthesis*, Island Press, Washington, DC

Meadows, D., Meadows, D., Randers, J., Behrens, J. and Behrens, W. (1972) *Limits to Growth: A Report to the Club of Rome Project on the Predicament of Mankind*, Universe Books, New York

Mensah-Kutin, R. (2008) *The Intersection of Gender, Climate Change and Human Security in Ghana*, ABANTU for Development, Accra

Merchant, C. (1980) *The Death of Nature:Women, Ecology and the Scientific Revolution*, HarperCollins, New York

Momsen, J. H. (2004) *Gender and Development*, Routledge, London

Monimart, M. (1989) 'Women in the fight against desertification', in *IIED Paper*, International Institute for Environment and Development (IIED), London

Moser, C. (1993) *Gender Planning and Development: Theory, Practice and Training*, Routledge, New York

Mukhopadhyay, D. (2008) 'Indigenous knowledge and sustainable natural resource management in the Indian desert', in C. Lee and T. Schaaf (eds) *The Future of Drylands*, UNESCO and Springer, Paris/The Hague, pp161–170

Neumayer, E. and Plümper, T. (2007) *The Gendered Nature of Natural Disasters: The Impact of Catastrophic Events on the Gender Gap in Life Expectancy, 1981–2002*, London School of Economics, University of Essex and Max-Planck Institute of Economics, London

O'Brien, K. (2007) 'Commentary to the paper of Úrsula Oswald Spring: Climate Change: A gender perspective on human and state security approaches to global security', paper presented at International Women Leaders Global Security Summit, 15–17 November 2007, New York

Ortner, S. (1974) 'Is female to male as nature to culture?', in M. Z. Rosaldo and L. Lamphere (eds) *Women, Culture and Society*, Stanford University Press, Stanford, CA, pp68–87

Ortner, S. (1990), 'Gender hegemonies', *Cultural Critique*, no 14, pp35–80

Rao, A. and Kelleher, D. (2005) 'Is there life after gender mainstreaming?', *Gender and Development*, vol 13, no 2, pp57–69

Rocheleau, D. E. (1985) 'Women, environment and development: A question of priorities for sustainable rural development', in *Background Paper Doc.No.B3/E*, ICRAF, Nairobi

Seager, J. and Hartmann, B. (2005) *Mainstreaming Gender in Environmental Assessment and Early Warning*, UNEP, Nairobi

Shiva, V. (1988) *Staying Alive: Women, Ecology and Development*, Zed Books, London

UN (1999) *1999 World Survey on the Role of Women in Development*, UN, New York

UNDP (United Nations Development Programme) (2005) *Human Development Report 2005. International Cooperation at a Crossroads: Aid, Trade and Security in an Unequal World*, UNDP, New York

Vainio-Matilla, V. (2001), *Navigating Gender: A Framework and Tool for Participatory Development*, Ministry of Foreign Affairs, Helsinki

van Wijk-Sijbesma, C. (1985), *Participation of Women in Water Supply and Sanitation: Roles and Realities*, IRC, The Hague

van Zon, H. (2002) *Geschiedenis en Duurzame Ontwikkeling, Vakreview Duurzame Ontwikkeling [History and Sustainable Development. Disciplinary Review Sustainable Development]*, DHO, UCM/KUN, Nijmegen, Amsterdam

Ward, B. and Dubos, R. (1972) *Only One Earth: The Care and Maintenance of a Small Planet*, Penguin, Harmondsworth

WEDO (Women's Environment & Development Organization) (2007) *Changing the Climate: Why Women's Perspectives Matter*, factsheet, WEDO, New York

Part I
The Analysis

2

Gender, Environment and Climate Change: Understanding the Linkages

Irene Dankelman and Willy Jansen

This chapter looks into the dynamic relationships between gender and environment, as manifestations of the interconnectedness between the socio- and the eco-sphere. It will introduce some theoretical reflections on these linkages, and give insight into major trends and developments since the 1980s. Implications for understanding gender–climate change nexuses will be highlighted throughout this chapter and will be explored in more detail in the chapters that follow.

Introduction

In order to understand gender-specific causes and impacts of climatic variability and changes, as well as related coping and adaptation strategies, we will first look into human dimensions of natural resource use and management. In order to understand the importance of the gender dimensions in environmental management and knowledge development, a thorough assessment of the gender-specific aspects of environmental conditions is necessary.

That environmental use and management and social relationships are closely linked is clearly demonstrated in the daily lives of millions of local women and men. It is in their lives and livelihoods that the eco-sphere and socio-sphere interact with each other. These interactions are gender-specific.

Over the past two decades, much has been written about the gender-specific roles in environmental use and management. As many studies and publications have indicated, women and men worldwide play diverse roles in the use and management of natural resources at a local level and carry diverse responsibilities and rights, including access to and control over natural resources.

In her article on ecological transitions and the changing context of women's work in tribal India, Menon (1991) introduced the concept of 'work' as a determining factor for gender dimensions in environment. She describes 'work' as active, labour-based interactions of human beings with the material world. Historically this interaction has been intricately based upon the natural environments in which communities lived and survived, since nature and natural resources and processes represented the material world on which human survival and well-being depended (Owen, 1998). Such resource-dependent livelihoods are still abundant throughout the world, not only in rural but also in urban settings (Slater and Twyman, 2003) (see Chapter 4). Menon distinguishes the following major areas of human work in those communities: food procurement, including food gathering and production; the protection of life, property and territory, including the collection of water, energy sources and fodder; and childbearing and rearing, including the maintenance of basic health standards and collection of medicinal plants. In most of these tasks, women play a predominant role and in performing these they interact directly with the natural environment. Important categories are production, reproduction and distribution of resources and rights. Based on these interactions Davidson et al (1992) describe the specific roles of women in local communities as providers of 'primary environmental care'. Already in her book *Male and Female* of 1949 (reprint 2001), Margaret Mead underlined that traditional tribal economies show a division of labour along gender lines. Many of the gender specific tasks and responsibilities are directly interfering with and dependent on the natural environment. Gender-specific divisions of labour related to the environment are still visible in modern societies: women and men perform different tasks and carry diverse responsibilities. However, in many cases women are disadvantaged compared to men with regard to land and water rights, and rights over other (natural) resources. Their access to and control over such resources are often dependent on their spouses and their relationships with other males in the community. The decision-making power of local women is also limited in comparison to that of men. So it is in their divisions of labour, responsibilities and rights that human relationships with the environment and natural resources are gendered, and power relationships are important determinants in that respect. Such situations are not static but change over time depending on circumstances at local, meso- and macro-level.

Authors from the eco-feminist school assign an intrinsic and emotional relationship between women and the environment. Eco-feminism emerged in the 1970s and 1980s. The term was introduced by Françoise d'Eaubonne in her book *Le Feminisme ou la Mort*, published in 1974. Eco-feminists, such as Carolyn Merchant (1980) argue that women are closer to nature than men, while men are closer to culture. Many eco-feminists see a connection between male domination of nature and male domination over women, stemming from the period of so-called scientific and cultural Enlightenment. Vandana Shiva, Indian physicist and activist, underlined in her book *Staying Alive: Women, Ecology and Development* (1988) that paternalistic, colonial and neo-colonial forces and values very often have marginalized women's knowledge. It is male dominated 'mal-development' – often assigned to 'white' men – which has caused major social and environmental problems. Although these alignments are based on the interaction that is reflected in women's and men's daily work in the natural environment, also other social roles – such as those of caretaker, childbearer and raiser – are ascribed as major contributors to the gender-specific roles of women and men in relation to natural resources and the environment.

The assumption that women are closer to nature and men closer to culture was not only defended by eco-feminists. Ortner (1974) recognized this association as a pervasive ideology in many societies and criticized it not only as essentialist, but also saw it as the ideological underpinning of gender inequality as in many of these societies culture was valued over nature, and men thus over women. Later critics of eco-feminism, such as Braidotti (1993) and Agarwal (1998) underline that eco-feminists primarily have focused on ideological, essentialist arguments and have failed to address power and economic differences as important sources of dominance. They argue that many eco-feminists do not differentiate women themselves by class, ethnicity and caste, nor recognize that concepts of nature, culture and gender vary across different cultures and localities. Agarwal (1998) suggests an alternative concept, namely that of 'feminist environmentalism'. Her approach is similar to what Rocheleau et al (1996) have called 'feminist political ecology'. This concept insists that the link between women and the environment should be seen as structured and shaped by a given gender and class/caste/race organization of production, reproduction and distribution. According to their views, the class-gender effects of environmental change are manifested as pressures on women's time, income, nutrition and health, social support networks and knowledge (Agarwal, 1998, 2000). In the next section such elements and relationships will be highlighted.

Enhancing food security

In many sectors, including the environmental, women's roles and responsibilities as food providers are frequently overlooked and underestimated. According to several researchers and writers, such as Dahlberg (1983), Owen (1998) and Howard (2003), it was actually 'woman-the-gatherer' who was the primary source of sustained food supply for local prehistoric communities – and not 'man-the-hunter'. Women's activities – among which were the gathering of fruits, nuts, edible leaves, flowers, mushrooms, roots, shoots, tubers, biomass energy and medicinal plants – provided daily sustenance. Meat was merely a supplementary food item, except for in the Arctic region where it was much more important (Lee and De Vore, 1968).

Women in many communities around the world still play an important role in biodiversity use for household sustenance through food collection. For example, the Brahui women in the Noza sub-watershed in Baluchistan, Pakistan, go out in early spring after the cold winters (FAO, 1997). Walking in groups or sometimes alone, they collect tiny edible plants and mushrooms. The plants are green and succulent and all lumped together as 'spinach' even though each plant has its own name. These spring greens are sometimes life-saving and provide much needed nourishment after a long winter of food deficits. When medicinal plants are flowering between May and June, small groups of women walk in the early mornings to collect these plants. In a Food and Agriculture Organization of the United Nations (FAO) supported project in the area, women participating in the walks expressed their desire to record their collected information and knowledge of plants. Zer Malik, one of the women, said: 'We want our daughters to be able to see how much knowledge their illiterate mothers actually possess'(FAO, 1997, p6).

Women dealing with wild seeds and vegetable crops have played important roles in the revolutionary innovation from gathering into the production of food. Because food collection requires a thorough knowledge of plant growth and animal life, maturation and fruition or reproduction, women have been credited with the discovery, domestication and cultivation of plants and animals and the intervention in selective breeding. According to several scholars they discovered the propagation of shoots and cuttings, seed selection and the construction of seed beds (Murdock and White, 1969; Stanley, 1981; Owen, 1998).

Stanley (1981) has credited the following interventions in cultivation to women: the use of ash as fertilizer; the creation of work tools such as the hoe, spade, shovel and simple plough; fallowing and crop rotation; mulching, terracing, contour planting, irrigation and land recuperation through tree planting. According to the same author, eight of the most important cereals worldwide are all domesticated by women: wheat, rice, maize, barley, oats, sorghum, millet and rye. Sir Albert Howard underlines in *An Agricultural Testament* of 1940, that he saw in India's peasants a knowledge of farming far more advanced than that of the West. In that land-use women played an essential role.

In studies on women's roles in food production in villages in the Garhwal Himalaya in India (Shiva et al, 1990; Kelkar and Tshering, 2004; Meuffels, 2006), it became clear that women are important actors in local farming that is based on sustainable flows of fertility from forests and farm animals to croplands. These century-old food production systems have always included and integrated forests as well as farm animals in the crop cultivation process. The internally recycled resources provide the necessary inputs for seeds, soil moisture, soil nutrients and pest control.

Singh (1988) described women's extensive contribution in animal husbandry in Northern India as follows: the woman harvests the crops and stakes the hay for domestic animals. She transports the leaf fodder and bedding material over long distances on difficult terrain. She grazes the cattle on distant grazing lands, brings animals to water sources, takes care of the young calves, milks the animals and cleans the animal shed. Particularly the collection of fodder leaves, herbs and grasses is almost exclusively a women's task, assisted by children, often girls. Male responsibilities are mainly in the ploughing, castration, purchase and sale of farm animals.

Next to men, women also work in irrigated agriculture: a Grameen Krishi Foundation project in northwest Bangladesh showed that women carry out about 50 per cent of all tasks in rice production. They even account for 50 per cent in the presumably male task of irrigation (Jordans and Zwarteveen, 1997). However, women's rights to water for agriculture vary enormously. In the Andes, women were allowed to participate in the construction of irrigation systems and thus to establish rights to irrigation water. But men dominated the written registration processes and decision-making bodies (NEDA, 1997). In Tanzania, by contrast, women were prohibited from operating water infrastructure facilities (NEDA, 1997).

According to the FAO (2009), rural women are the main producers of the world's staple crops – for example in southeast Asia they provide 90 per cent of labour in rice cultivation and in Pakistan 50 per cent of rural women cultivate wheat. Their contribution to secondary crop production, such as vegetables, is even greater. Women make up 51 per cent of the total agricultural labour force. This includes work in subsistence farming as well as export-oriented farms and plantations (FAO, 2009). In Burkina Faso, 95 per cent of women work in subsistence farming and in the informal sector. In El Salvador the gender division of labour differs according to crops and type of activity: male farmers are primarily responsible for cash cropping, whereas women are responsible for food production, especially basic grains, vegetables and fruits (FAO, 2009). Another characteristic is that in many countries women work longer hours than men, and earn less

(UNFPA, 2002). Singh (1987) accounts that in a year a pair of bullocks works for 1064 hours on a 1ha farm in the Indian Himalayas, a man for 1212 hours and a woman for 3485 hours. In the Noza sub-watershed in Pakistan, an average working day for a Brahui woman is 17 hours long during the productive season (FAO, 1997, p3). In four countries in sub-Saharan Africa – Burkina Faso, Nigeria, Kenya and Zambia – women's average daily time investment in agricultural work was 467 minutes a day, whereas men's time investment accounted for 371 minutes a day (Saito et al, 1994, p18).

Through their daily work, women have developed a profound know-ledge of the wild plants and animals used, as well as the indigenous varieties of crops. For example, women rice growers in central Liberia, India, identified 100 indigenous rice varieties with two or three errors at most, not only describing the different varieties, but also mentioning other features, such as the ease with which the husk can be removed, the length of time required to cook, and their suitability for different ecological conditions (Shiva and Dankelman, 1992, p46). In a small sample participatory study with women hill farmers in Dehra Dun, India, Shiva was provided with no less than 145 species of forest plants that women have knowledge of and which they utilize. External forestry experts could only name 24 species (Shiva and Dankelman, 1992, p46). The Brahui women in the Noza sub-watershed, Pakistan, were able to identify 35 medicinal plants during field walks (FAO, 1997, p10). In the book *Women & Plants: Gender Relations in Biodiversity Management and Conservation* (2003), edited by Patricia Howard, numerous cases are presented of women's plant knowledge and involvement in plant use and conservation worldwide; in Mexico, Venezuela, North America, Zimbabwe, Swaziland, Mali, Turkey, Italy, Nepal, Thailand, China and Papua New Guinea.

Also in fishing communities there is a strong gender division of labour and next to men women play an important role. Although fishing is predominantly a male occupation that exposes them to dangerous situations at sea or other water bodies, occasionally women go out fishing themselves. More often they handle the preservation and marketing of fish (Oracion, 2001). Between 50–70 per cent of pre and post-harvest activities in fishing communities are done by women, and their per capita income derived from fish marketing and processing is in fact often higher than the per capita income of their husbands and sons from fish capture (Nozawa, 1998). Fisherwomen are mainly responsible for the preparations of the fishing trip – including the mending and boat building – fish processing and selling, while men often do the actual fishing. Water and firewood hauling, necessary for fish drying, are also mainly carried out by women (Aguilar and Castañeda, 2001; Williams et al, 2006). Other

Box 2.1 *Weather, climate and food production*

Local conditions, particularly weather and climatic circumstances, determine to a large extend the productivity of land and water resources. Not only are these therefore important determinants of people's food security, but they impact also on women's and men's work, tasks and responsibilities in those systems. Water availability and precipitation, salinity, air temperatures, and wind currents determine if the weather and climate is favourable for food production. Drought, caused by absence of rainfall and high temperatures (evaporation), can cause failure of harvests and result in famine. On the other hand, flooding can be important for the replenishment of nutrients for crop production, however, too much flooding causes damage and destruction of crops and harvests. Not only the mean temperature of the air but also extremes such as freezing, and periodicity, determine crop production. Such conditions not only influence plant life, but also that of cattle, other animals and disease vectors. Water temperatures and composition are important for life forms in the water and therefore for the health of coral reefs and fishes, determining fishing opportunities. All these conditions can have a positive and negative influence on women's work as food producers, and the lives and health of local communities.

water organisms such as shellfish and prawns are often being collected and managed by women (Pijnappels, 2006). However, their decision making power and access to technical training and development assistance in the sector is limited. The Second Global Symposium on Gender and Fisheries of the Asian Fisheries Society that took place in 2007, offered a solid base for understanding gender differentiated roles and development needs in fisheries and aquaculture. Several research papers reported that enhanced women's contributions add value to the sector (Williams et al, 2006; Choo et al, 2008).

Household chores: Reproductive tasks

Because of the fact that one of the major areas of human work, childbearing and rearing, is predominantly assigned to women, Menon (1991) and others argue that not only tasks such as the provision of health services and hygienic measures, but also related aspects, such as the provision of household energy and water, are primary among women's responsibilities. This is not only true in rural, but often also in urban situations (see Chapter 4).

In many societies, women and often girls are exclusive suppliers of water for household use: they collect every litre of water for cooking, bathing, cleaning, maintaining health and hygiene. They also play a predominant role in the provision of water for animals, crop growing and food processing. It is often women who decide where to collect water, how to draw, transport and store it, what water sources should be used for which purposes and how to purify drinking water. They make a disproportionately high contribution to the provision of water for family consumption compared to male members of households (GWA, 2003). In segregated communities, where women are not allowed to be seen in public, daughters often bear the burden of water collection (NEDA, 1997). Male family members rarely assist in the often heavy and time-consuming task of water transportation for family use, and usually only then if they have bicycles or carts. Through their work, women have acquired specialized knowledge in the area of local water management and use. It is a knowledge they primarily share with their daughters and each other. Because of other tasks they perform, such as the washing of clothes and dishes, cleaning houses and latrines, and attending to personal hygiene, in several regions women have established specific ways for reusing waste water to conserve supplies (Dankelman and Davidson, 1988; GWA, 2003). Sometimes women's needs for water can be in direct conflict with those of men: for example, food production can be an important source of family food and women's income, but women's access to irrigation for such production is minimal as this sector is dominated by male members of the household (UNDP, 2002).

Most domestic energy comes directly from biomass sources. Wood fuels – both in the form of firewood and charcoal – and other biofuels, such as animal and crop residues, form the only source of energy for about 2 billion people, and some 1.5–2 billion have no access to electricity. Globally fuelwood production was at about 1.8 billion cubic metres in 2005, an increase of more than 6 per cent since 1990 and of 30 per cent since 1970 (FAO, 2007). Today more wood is used for fuel than for any other purpose and the requirement for fuelwood – 5 per cent of global energy consumption – is increasing rather than decreasing (Forsyth, 2005, p260). In urban areas there is a transition from high wood use to fuels such as gas and electricity, with charcoal as the principal fuel that people shift to first from fuelwood (CIFOR, 2003; Arnold et al, 2003). Poor households, even in cities, still depend almost exclusively on biomass sources for their energy supply, but they are often least likely to have fair access to the resources (Arnold et al, 2003). Although men sometimes share the task of household energy supply – particularly if resources are commercialized – in many communities women have the primary responsibility for meeting household energy needs through fuel collection, preparation (such as chopping and drying) and use, including tending the fire and cooking. All these tasks may take many hours per day, depending on the ecological circumstances and access to biomass resources.

The majority of rural women in Asia, Africa and Latin America obtain fuel wood, food and fodder from nearby forests and woodlots. Children's role is often an extension of their mothers' work, assisting with fuel collection, fuel preparation, cooking and fire-tending, the latter mainly by female children (Cecelski, 1985; Forsyth, 2005). Men in the Garhwal hills of the Himalayas, India, are found to break the traditional division of labour only by fetching fuel and fodder when the productivity of women's labour is high, for example on irrigated land. When domestic fuel becomes more commercialized and collection is oriented towards large-scale organized sale and charcoal making, men's participation increases. But as long as technology and marketing are limited or absent, the task of fuel gathering is regulated by women. Often they carry loads up to 35kg or more over distances as much as 10km from home. The weight largely exceeds the maximum weight of 20kg permissible for women by law in many countries (ILO, 1967; WHO, 2004, 2006). Indoor air pollution from biomass fuel and exposure to domestic smoke while cooking, causes major health problems, including chronic respiratory disorders, particularly in women (Chen et al, 1990; WHO, 2004, 2006).

In many so-called traditional societies, women have played and continue to play an important role alongside men in the construction and management of human shelters and infrastructure (Steady, 1993; de Vries and Keuzenkamp, 2000). Households closely reflect the conditions of the surrounding physical environment and it is women, often assisted by female children, who bear the responsibility for protecting members of their households, especially the young and aged, from pollution, poor sanitation, natural disasters and the risks inherent to poor housing conditions (WHO, 2004). Women may spend as many as 20 hours a day at home, especially in secluded societies. In many cases, activities in human shelters and physical infrastructure come to depend on women's unpaid labour. The responsibility of maintaining a clean and safe household environment, including waste management, still falls primarily on women's shoulders. Poor women are also engaged in activities such as waste picking from dump sites and waste salvaging tasks (Muller and Schienberg, 1998; Bulle, 1999).

Income generation: Productive work

Women made up 40.5 per cent of the global labour force in 2008 (ILO, 2009a, p9). Gender inequality remains an issue within the labour market globally, as women suffer multiple disadvantages in terms of

Box 2.2 *Weather conditions and access to water*
and other natural resources

Precipitation, its volume and its seasonality – together with other physical and vegetation conditions – determine to a great extent if people are living in arid or more wet circumstances. Access to water of good quality and the absence of extreme weather events impact on women's work in providing water for the household or irrigation purposes. Water quality, including its salinity, also determines to a great extent whether the available water can be used. Salinity can be determined by high evaporation and influx of sea water.

Productivity of vegetation, on the other hand, depends on soil, water and weather conditions; this is not only true for food production but of course also for natural vegetation, including pasture and forest growth. In turn the (adequate) availability of natural resources and non-timber forest products (NTFPs) – for biomass energy, fodder for livestock, medicinal and other uses – impacts on women's work and that of their children. Even for the production of electricity, including hydroelectricity, abundant water is used. Particularly smallholders, subsistence farmers, pastoralists and artisanal fisherfolk are likely to suffer complex and localized impacts of climate change (IPCC WG2, 2007).

access to labour markets and freedom to choose to work. Throughout the developing world, vulnerable employment, such as own-account workers or contributing family members, is generally a larger source of employment for women (52.7 per cent) than for men (ILO, 2009a, p11). A majority of women in the developing world (except for northern Africa) are disproportionally represented in informal employment – outside agriculture, as street vendors, waste collectors, home-based producers, garment workers and domestic workers (Chen et al, 2005; Chant and Pedewell, 2008; WIEGO, 2009). Much of this work is hidden and does not show up in statistics because it is illegal – such as prostitution – or culturally not categorized as income-generating work (Jansen, 2004). In sub-Saharan Africa and South Asia the agricultural sector makes up more than 60 per cent of all female employment (ILO, 2009a, p9). Women are entering the global formal workforce in record numbers, but they still face higher unemployment rates and lower wages and represent the majority of the 1.2 billion poor workers, and 630 million workers who are living in extreme poverty (ILO, 2009b). The conditions under which many women work and the access women have to employment and productive resources, can differ considerably from men's, with lack of safety nets, inadequate

Box 2.3 *Weather, climate and employment*

Many income-generating activities in which women and men are involved, particularly those in the informal sectors, are indirectly or directly dependent on the availability of raw materials, energy and water resources. The dependency of the agricultural sector has already been described above. This means that all food-based processing and trading sectors, in which many women earn their incomes, depend on weather conditions. Similarly, activities such as brick making and artisan work, depend directly on the availability of raw materials, such as NTFPs and biomass energy sources. Many commercial estates, in which men and women find employment, use large amounts of water, energy and raw materials from agriculture or nature. In many cases, availability of these resources depends on weather and climatic conditions. For example, in an interview with women farmers in Guyana it was found that their pineapple growth and related industry was suffering from the prolonged droughts (WEDO/UNIFEM, 2010).

remuneration, unsafe conditions and disrespect of fundamental human rights (UNFPA, 2006; ILO, 2009a). Informal work has expanded and appeared in new forms in the context of globalization, neo-liberalism and cross-border and rural–urban migration (Chant and Pedwell, 2008).

Many of the informal income generation activities in which women are involved are directly or indirectly dependent on natural resources, such as energy sources, NTFPs, crops and water. Activities include, for example, agricultural and plantation work, processing and selling of food products, brick making, handicrafts, pottery, spinning, weaving and sewing (Jansen, 2007). Head loading of fuel wood for sale in urban areas and the preparation and marketing of charcoal are important income-generating activities for women in certain regions. In industries such as leather tanning, workers – including many women – are directly exposed to the environmental conditions of water and air. Occupational health issues are often a major concern, particularly for the health situation of poor women workers worldwide (WHO, 2004).

Analysis: Sustainable management and use of resources

In all activities described above, men and women play an important role by inputting their work, energy and expertise. Through these activities women

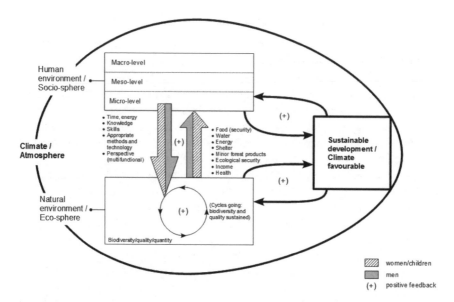

Source: Adapted from Dankelman (2007); technical design Jeroen Dankelman

Figure 2.1 *A sustainable system: Gender aspects of natural resources management and climate (change)*

contribute substantially to their families' security of food, health, energy, shelter, production and income. All these activities depend on natural resources, the physical environment and ecological functions and services. Rural women's work in particular is therefore often directly related to the natural environment and environmental conditions. Although men also perform tasks in the above-mentioned fields, their input of time and energy is often substantially less than that of women. In this way, rural women in many countries maintain to a large extent – alongside men – the interface between the socio-sphere and eco-sphere.

In Figure 2.1, an environmentally sustainable system is presented. As has been shown above, this does not automatically mean that the system is socially just and prosperous; even in these situations poverty levels can be high and inequality between and within the sexes and different classes, castes, ethnicities and ages can prevail. This is certainly the case for the growing number of female-headed households.

In the figure, a distinction has been made between the human environment or the socio-sphere, and the natural environment, or the eco-sphere. In the socio-sphere, different levels are distinguished to differentiate between the processes taking place at macro-level – for example at international level, at meso-level (such as national policies) and at micro-level – in the realities where local people actually live. All these levels interact with each other, and within each level, social interactions take place based on local circumstances and power dynamics. As Leach (1992) and Rocheleau (1995) have underlined, women's roles in these interactions are not fixed and should not be generalized.

In a healthy eco-sphere, an abundance of natural resources and large diversity of plant and animal species and varieties, as well as (agro-) ecosystems are present. The physical environment – including water, soils and air – has a quality and quantity that sustains diverse life forms, including human life. The components of the eco-sphere are related to and interact through ecological cycles. Both the socio-sphere and eco-sphere show a dynamic balance within themselves and with each other. Local people contribute energy, time, knowledge and skills, and cultivation and management practices – as well as technologies to manage the eco-sphere, or livelihood system. This way they yield from the system what is necessary for the family's and community's subsistence on the short and longer term, such as food, water, energy, NTFPs, medicines and shelter. The ecosystems provide a diversity of goods, and regulating and supporting ecosystem services contribute to human health, well-being and livelihood (UN Millennium Project, 2005; Millennium Ecosystem Assessment, 2003).

This way, one could speak about an ecologically sustainable system: there exists a positive interaction between the different factors, components and levels of the eco-sphere and socio-sphere, and there is a dynamic balance between that which is used from the (agro-)ecosystem, that which is provided by it, and its regeneration (or buffer) capacity. Because of their tasks, work, knowledge and expertise, local women, men and even children play an important role in maintaining that balance.

Conditions and critical factors

Many factors have an influence on these environment-related tasks of women, men, and children, and therefore have an impact on their work-burden, physical and psychological stress and autonomy. Apart from the division of labour, tasks and responsibilities, critical factors are:

- Access to and control over (natural) resources of good quality, such as land, water and trees.
- Access to and control over other means of production, such as income, credit, and appropriate technologies.
- Access to education and training.
- Social status and decision-making power, regarding resource management and use, production and produce.

- Participation and involvement in social processes and freedom of organization.

These critical factors are not only essential at micro-level within households and communities, but also at meso and macro-levels.

Apart from economic, cultural and technological circumstances and power structures, environmental conditions also have their influence on women's and men's work and lives.

This is true most directly in rural situations, but these aspects can also be observed in urban environments. Particularly when it concerns people who live in poverty and who are marginalized because they depend to a large extend on freely available resources (see Chapter 4).

People facing environmental problems

All over the world, local communities face problems of environmental degradation and deterioration. These problems are as old as humanity, but have increased over time in magnitude and intensity. Seager (1993) argues that an understanding of these environmental problems needs to be rooted in an analysis of the social, cultural and political institutions that are responsible for environmental distress. Like Shiva (1988), she underlines that the institutional culture responsible for most of the environmental calamities in the last century is primarily a masculinist culture. Of course both men and women have participated in this culture, but women's voices and decision-making power have been more limited and restricted.

Often developments such as commercialization, export-orientation, international trade and prizing policies, external debt, structural adjustment programmes, political and armed conflicts, as well as increasing consumption and population pressures have resulted in over-exploitation or pollution of natural resources. Such processes are characterized by redirection of uses of community land and other resources (including forests, grazing grounds and water sources), extraction activities (including logging and mining) and introduction of non-sustainable agricultural and industrial processes and infrastructure. This results in scarcity of common resources, degradation of their quality and diversity, and disruption of ecological functions, such as retaining and pollution of water, soil and air (Hardin, 1968; Goldman, 1997; Radkau, 2008). It is often women – and with them children – whose lives and positions are the hardest hit by such developments.

Figure 2.2 shows what happens when a system becomes unsustainable, both at the level of the eco-sphere and the socio-sphere. At the level of the eco-sphere, more resources are taken than can be regenerated, resulting in a decline in biodiversity. The ecological cycles in the system become destabilized – ecological processes and services involving natural resources, such as water, soil and biodiversity, get disturbed. More pollution is added to the system than it can handle; in other words, the ecological quality diminishes.

At the same time it is observed that the socio-sphere becomes more destabilized. A degrading environment is an important driver in the increase of poverty and inequity. This is also reflected in changing power relationships in society, community and family, among genders. Women's access to and control over natural resources and technology generally decreases more than that of male members of households. Women often lose control over resources, production and management, and become more marginalized and excluded.

In these situations, which are becoming more and more common, community members, especially women, have to put more time, energy and effort into meeting their families' basic needs for natural resources, security, and health. However, as the eco-sphere cannot supply enough and the natural cycles that sustain life are disturbed, this task becomes more difficult and sometimes even impossible. Because women have to walk longer distances, often over rough or disturbed terrain, women's work and life becomes more insecure. In Sri Lanka, it was found that women spend

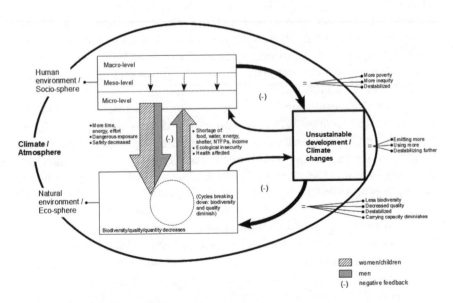

Figure 2.2 *An unsustainable system: Gender aspects of natural resources management and climate (change)*

Source: Adapted from Dankelman (2007), technical design Jeroen Dankelman

Box 2.4 *Climate change impacts on people's livelihoods*

As climate change manifests itself in weather extremes and disasters, casualties in the lives of women and men, injuries and health impacts are direct manifestations of these problems. As has been widely observed, climatic changes not only impact directly on people's lives, but also affect the security of their livelihoods in all its forms, including their food security, water security (see Box 2.5), energy security, security of shelter and ecological security. In all these aspects, impacts for women and men are different, with the most vulnerable suffering the severest impact. Finally, climate change impacts differently on the dignity of women and men, jeopardizing their human rights (such as the increase of violence against women because of increased stress factors), lack of education and training opportunities, and active participation and involvement in policies and actions (see Chapter 3).

9–15 hours per week gathering fuelwood, over distances ranging from 0.5–3.5km, carrying bundles weighing 28–34kg per trip (Wickramasinghe, 1994, p32). In Baluchistan, Pakistan, each woman spends an average of 15 hours per week on fuelwood collection in summer and even more in the winter (Imran, 2002, p57). In Northern Kenya women gather firewood about three days per week. The average weekly time spent on fuelwood gathering was 8.2 hours, but in the more arid sites up to 10.5 hours per week were spent with loads up to 45kg per trip (McPeak, 2002, pp7–8). A study executed in Nepal learned that a degrading environment, with the forest cover going down by more than half in the past 20 years and accompanying erosion, landslides and flooding, particularly affected girls' lives and opportunities. Whereas a myriad of development programmes had been initiated to enable more girls to attend primary schools, the scarcity of natural resources and the time and energy needed in collecting and managing these, even resulted in a decrease in girls' school enrolment (Johnson et al, 1995).

> *You people in the cities do not know how village women like us have to slog. As the farmer forces his bullocks to labour, poverty forces us to labour all the time. One side is getting inflated while the other is shrinking. The section that is shrinking is getting destroyed.*
> (peasant woman from Bangladesh; Mazumdar, 1994, p17)

Climatic changes are not only hitting women, men are also forced to adapt. Malin Jennings of the Arctic Indigenous Climate Change Ethnographies

(ICCE) underlines that the warmer climate has made survival harder particularly for male Inuit hunters in Greenland: 'Hunters used to be the pinnacle of society but this is no longer the case. Their skills and expertise have been rendered useless, and this is creating social problems' (Jennings, in Parbring, 2009, p1).

A major alternative strategy primarily for men under circumstances of severe resource scarcity is migration. With resources already being scarce, women in the communities have to take over production tasks from their husbands. In such cases they may have to rely on wage labour, apart from other tasks and responsibilities, resulting in a tremendous pressure. In Diourbel, Senegal, male out-migration was a logical response to the modernization of agriculture and declining food security. While women in the region are concerned about the changing environment around them, they are in a weaker position to do anything about it (David, 1995).

Migration puts also men under pressure, as they have to find sources of income quickly to make up for losses or insufficiencies in livelihood, and are compelled to live away from their families (Leduc, 2009; Hunter and David, 2009). 'I don't want to leave this place,' said Gaurpodomando, a fisherman in Harinagar, Bangladesh, suffering from the declines in catch year after year. 'I don't want to leave this country. I love this place' (in Friedman, 2009).

In a study of three villages in Burkina Faso, it was found that wild plants made up 21 per cent of the diet and were an important source of Vitamin A. The work of collecting these plants in the forests, which are 8–10km from the villages, is 85 per cent done by women and young girls. Seasonality is an important factor in women's time use, but also environmental degradation. As time taken to obtain these wild plants increases due to environmental change, the wild foods are used less and less, causing dietary deficiencies, particularly in families with inadequate cash resources to supplement their diets (Awumbila and Momsen, 1995, p338).

The use of toxic chemicals in agriculture poses serious health threats for labourers, particularly women. Weeding is predominantly a women's task. Extensive use of pesticides in agriculture is posing major health problems for sprayers, producers and packers. In Malaysia, many women work in the plantation sector and spray pesticides. They complain of sore eyes, skin rashes, burnt fingernails and disruption of the menstrual cycle. Some mothers exposed to pesticides during early pregnancy give birth to deformed children or even lose their unborn children (Arumugam, 1992; UBINIG, 2003).

We know that pesticide is harmful. When we mix the pesticide in the water it expands like boiling milk. It reacts in our body in the same way. We observe that birds and animals are dying. We women do not want to use pesticides because we love our chicken, cows and goats like our family. (Jamila Khatun of Raini Karmakar para village in Bangladesh, in UBINIG, 2003, p12)

He cries all the time. I can't go out to work. It would be impossible to have a job and take care of my son. We have to make do with the tiny amount of money (we have). (Jayanthi, mother of Harshith, a boy from India who has become mentally and physically challenged because of aerial spraying of endosulfan, in Rengam et al, 2007, p65).

According to the UN, the number of people without improved drinking water has dropped below 1 billion, whereas 2.5 billion people lack access to basic sanitation (UNICEF/WHO, 2008, p2). Although the world seems to be ahead of schedule in meeting the 2015 drinking water target of the Millennium Development Goals (MDGs), there are still 884 million people worldwide who rely on unimproved water sources for their drinking, cooking, bathing and other domestic activities; of these, 84 per cent live in rural areas (UN, 2009, p48). About 80 per cent of those lacking access to drinking water live in sub-Sahara Africa, eastern and southern Asia.

According to the UN: 'Imbalances between availability and demand, the degradation of groundwater and surface water quality, intersectoral competition, interregional and international conflicts, all bring water issues to the fore' (UN-Water, 2006, p2). Today agriculture accounts for 70 per cent of all water-use globally, and up to 95 per cent in several developing countries (UN-Water, 2006, p6). Water use increased sixfold during the 20th century, which is more than twice the rate of population growth (UN-WAPP, 2006, p173). It is estimated that about two thirds of the world's population, or about 5.5 billion people, will face moderate or severe water stress in 2025 (UN-Water, 2006).

In many of the countries under water stress, women walk 1–4 hours per day over long distances to fetch water. In some mountainous regions in East Africa, for example, women spend up to 24 per cent of their caloric intake in collecting water (Lewis et al, 1994, p12). Despite their major responsibility in the field of water management, several factors restrict women's influence over this area of their lives. Male heads of household often decide where to build the family home, without necessarily considering the distance to water sources. In many cases, water rights are linked to land rights, restricting women's access to water sources as well. Often there are few or no public water points in slums and on the outskirts of cities, and in several regions there is a threat caused by privatization of water services

Box 2.5 *Climate change and access to water*

Climate change causes increased variability, exacerbating spatial and temporal variability of water resources. The serious deterioration of aquatic ecosystems contributes to these problems (UN-WAPP, 2006). The impacts of climate change on freshwater systems and their management are mainly due to the observed and projected increases in temperature, evaporation, sea level and precipitation variability. Semi-arid and arid areas are particularly exposed to the impacts of climatic change on freshwater, while higher water temperatures, increased precipitation intensity and longer periods of low flows are likely to exacerbate many forms of water pollution, with all its consequences for ecosystems and human health (IPCC WG2, 2007). Scarcity of clean water and unreliability of the supply will be major challenges for women and girls, as they are main suppliers of drinking water in many areas around the world.

(WEDO, 2003). In those instances, women have to collect the water from other sources, buy it from the provider or from vendors, whose prices can be high – making water an unaffordable resource.

Organic and inorganic contamination of water sources causes major health problems, particularly for those who are in direct contact with the polluted water – often women and children. In inner city areas, many of the poorest people live close to industrial areas, suffering high levels of pollution. Women predominate among urban people in poverty. Apart from lack of access to (legal) land, they face inadequate waste management, as well as water and sanitation problems. Many settlements of the poor are situated on dangerous sites that are vulnerable to landslides or flooding, or in the direct surroundings of hazardous industries. As pollution in many urban areas increases, the users and workers are more exposed to dangerous substances and toxic chemicals. This happens where housing sites are situated close to polluting and dangerous industrial estates, waste dumps or open drains and sewers. The Bhopal disaster of 1985 showed the disastrous effects of such situations on people, particularly women and children, living around the industrial plant of Union Carbide. Women told Menon (1991) that they had migrated to this industrial area because the households owned little or no agricultural land any more, fuelwood was getting scarce and life under such circumstances had become very hard. But now, living in such a polluted area, they face other severe problems.

While living in squatter areas and slums is bad for everybody and polluted air and water threaten the health of every inhabitant, women face great risks from exposure and handling of contaminated water, waste and other products, because of existing division of tasks and responsibilities. Also indoor air pollution poses a major problem for women's and children's health in particular. As the meal preparers, women are often exposed to high levels of smoke from open fires or stoves for long periods. The majority of 1.5 million deaths and of the 2.7 per cent of the global burden of disease from indoor air pollution every year occur among women (WHO, 2006).

Summarizing, as productivity of the ecosystem declines, shortages of basic supplies, such as food, water, energy sources and forest products occur. Livelihood security worsens and income generation possibilities diminish. Ecological insecurity increases and local communities suffer directly. The burdens on local communities, particularly on women and children, become heavier and their health and livelihood opportunities are threatened. Global warming contributes to the resurgence of diseases, such as malaria and Japanese encephalitis, and intensifies respiratory and water-borne diseases (Leduc, 2009). Ecological destabilization reinforces social dysfunctioning, poverty and inequality among classes, ages and sexes.

Pressure on the ecosystem increases, as more people are made dependent on less productive (eco)systems. This further diminishes the carrying capacity of the system, and reinforces environmental destabilization. In other words, the interaction between the actors and elements in the system becomes negative.

The main drivers behind environmental and related social destabilization is an unsustainable development trend, that takes more from both ecological and social systems than can be regenerated and is more destabilized than can be recovered (see Figure 2.2). A driving force behind unsustainable development trends is an economy based on short-term profits, stimulating increased and unequal production and consumption. These processes not only promote inequality between countries and regions, whereby the rich get richer and the poor lose their rights to and control over (productive) resources. Many of these development processes are not culturally adjusted.

In Figure 2.2, these external and internal processes are indicated by the pressure from the macro-level on the micro-level. Another aspect of unsustainable development is the introduction of inappropriate planning and technologies: these are not location-specific and are based on a constant flow of external inputs and expertise, as well as resource, energy and water intensity. Institutional – or governance – aspects also play a role: an unsustainable system often concentrates wealth and power, while marginalizing and excluding local communities, members of households (particularly women) and minority groups from decision-making processes. These processes do not impact every woman and man in the same way. It is important to not only recognize differences between women and men, but also to see the disparities between groups of different race, ethnicity, social status and age.

The convergence of global crises in recent years (such as the financial crisis), increasing commodity and energy prices at global markets, and environmental degradation and climate-related disasters, increase global hardship and insecurity. According to Sha Zukang (2009), UN Under-Secretary-General for Economic and Social Affairs, historically economic recessions have placed disproportionate burdens on women, as they are more likely to be in more vulnerable jobs, to be underemployed or without paid jobs, lack social protection and have limited access to and control over economic and financial resources. The UN Special Rapporteur on 'Violence against Women: Its Causes and Consequences', Yakin Ertürk (2009), stated that women and girls in developed and developing countries are particularly affected by these crises due to job cuts, loss of livelihoods, increased responsibilities in all spheres of their lives, and the increased risk of societal and domestic violence.

> *Climate change exacerbates existing inequalities and slows progress toward gender equality. Gender equality is a prerequisite for sustainable development and poverty reduction. But inequalities are magnified by climate change.* (Lorena Aguilar, International Union for Conservation of Nature, in World Bank, 2008, p42)

Critical factors

The above text indicates how unequal power relations, needs and perspectives, at all levels of society, fuel unsustainable and unjust developments. Consequences are reflected not only in the environmental field but also at a social level, in increased inequality and gender differentiation in use and management of natural resources, and in environmental change that adversely affects particularly women and children. The consequences of environmental degradation place extra burdens on women's – and girls'– shoulders, affecting their work, their health and their lives. Summarizing, critical factors in this respect are:

- Loss of access to and control over natural resources and (eco)systems of good quality, such as land, water, energy sources and minor forest products, but also seeds and biodiversity, particularly affecting women.
- Loss of women's access to and control over other sources of production, such as knowledge, technology, schooling and training.

- Loss of decision-making power, for example on joint resources management and common property.

Apart from these factors, for users an unsustainable (eco)system is characterized by:

- Increased need for time and energy input, and longer walking distances in order to meet basic family needs, resulting in overburdening and time poverty (World Bank, 2006).
- Increased effort needed to meet production needs.
- Increased shortages of basic resources – such as food, water and fuel – for day-to-day family needs and activities.
- Increased – and more direct – exposure to unsafe situations and dangerous substances.
- Breakdown of educational and income-generation opportunities due to lack of time, resources and labour.
- Restrictions on (women's) organization.
- Overuse of marginal resources – reinforcing the cycle of environmental degradation and poverty.
- Ecological and social insecurity, as well as vulnerability to natural disasters.

All this means that:

- Work burdens increase significantly.
- Health, survival and welfare are adversely affected.
- Poverty increases.
- Development opportunities become limited.
- People's autonomy and decision-making power declines.

All these factors have to be tackled in order to sustain people's livelihood systems and increase gender equality as they particularly affect women.

Coping strategies

Communities develop strategies to handle the problems described above. Many of these strategies depend on the local ecological, social and cultural context, and not all of these are sustainable themselves. Some of the coping strategies local women and men apply are:

- More time, effort and energy are put into work, particularly by local women. However, there are limits to how much time and effort one person can spend, particularly when this occurs over longer periods.
- Specific activities aimed at making available more natural resources and increasing their supply. Examples are women's initiatives in tree-planting and reforestation, as well as forest conservation activities. Kitchen gardens, installation of water points and regeneration of degraded land and watersheds, all through the active participation of women.
- People economize on the use of resources. A common strategy is, for example, shifting to other food products that need less cooking time (often these products are less nutritious), limiting the number of cooked meals or the boiling of water – with all its health consequences. Another possibility is the use of energy-saving or resource-saving devices. Many cases are known in which these technologies are introduced without adequately consulting women as users in their planning and implementation.
- Another issue, which has been taken up by some (groups of) women is reuse and recycling. In situations of water scarcity, for example, they manage to recycle and reuse water for several purposes.
- Communities also look into using alternatives, such as solar energy for cooking, switching to alternative crops or changing planting patterns or technologies. When the natural resource base becomes too limited to sustain livelihoods, a common strategy is to look into alternative means of income generation.
- Women in particular get organized. Already used to working together in the field or in the collection of natural resources, they share with each other the problems they face and look into solutions. Groups might be formed, or pre-existing women's organizations take up the environmental issues in their livelihoods.
- Prevention of pollution or cleaning up waste sites is another strategy local women and men use. Examples are waste-disposal activities, such as those started by a collective of unemployed women, such as in Bamako, Mali (Traoré et al, 2003).
- As consumers and producers, women can play powerful roles in the promotion of environmentally sound products and production processes. This is an issue throughout the world, particularly in high income countries, such as in Europe and North America.
- Local women and men organize to protest against environmental degradation and pollution, developments that threaten their resource base and livelihoods. In addition to holding protest demonstrations and campaigns, they often use non-violent means of opposition and blockades to stop such activities.

- In order to reflect their environmental and social concerns, many local women and men have created songs, poems and other forms of cultural expression. These can be powerful sources of inspiration in their struggles for sustenance of their livelihoods: 'Embrace our trees, save them from being felled. The property of our hills, save them from being looted.' (Chipko song, India)

Other actions needed

It should be strongly emphasized that it is certainly not the exclusive responsibility of local women, and in particular not of those that live in poverty, to change unsustainable livelihoods into more sustainable and socially just ones. Women should not (just) be seen as instruments for environmental regeneration and conservation or as passive recipients of development aid, but as equal partners in those processes. Environmental management is far more the direct responsibility of men and women in power at international, national and local levels, and of other development actors, including donors, the private sector and non-governmental organizations (NGOs).

Figure 2.3 indicates on which levels rectifying actions are presently being taken to restore sustainability, equality and justice.

These levels and interventions are as follows:

A. The natural environment or eco-sphere is preserved, in such ways as:

- Increasing the supply of natural resources, by reforestation, external inputs, and nature conservation.
- Regenerating the system and ecological cycles by land rehabilitation, erosion control, water management, ecological farming, multicropping and increase of biodiversity through management measures.
- Increasing the quality of the environment by the introduction of non-polluting processes and products, as well as waste and pollution treatment, sewage and sanitation.

B. Support to local communities, particularly women and children, by lightening their burdens and broadening their options, in such ways as:

- Introducing time and energy-saving devices.
- Developing vocational and natural management training and educational programmes, and promoting enrolment and participation.
- Increasing their access and control over production factors, for example by promoting changes in land tenure and legislation, access to credit and other financial facilities.

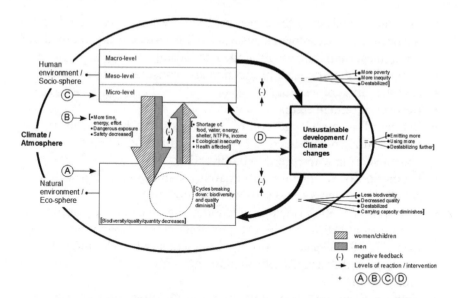

Figure 2.3 *Restoring sustainability: Gender aspects of natural resources management and climate (change)*

Source: Adapted from Dankelman (2007), technical design Jeroen Dankelman

- Promoting alternative income-generating and employment opportunities.

C. Promoting institutional changes at the socio-sphere level, in such ways as:

- Enhancement of participatory approaches and management systems, ensuring the effective participation of community members, particularly women.
- Institutional and legal changes to improve women's decision-making powers and rights to and organization of resources.
- Gender mainstreaming in organizations dealing with environmental issues and sustainable development.
- Promotion of environmental and social awareness at all levels of society.

Fundamental changes

D. As important as each of these activities are, they will remain a constant struggle as long as major drivers of unsustainable development are not tackled. Fundamental changes are needed at regional, national and international development levels, such as:

- Systematic analysis of the convergence of present crises, including the financial and environmental crises, as well as restructuring international economies, capital flows and trade relationships.
- Promotion of sustainable production and consumption processes.
- Redistribution of wealth.
- Respect of human rights.
- Increased access to and control over resources by local users and safeguarding of intellectual property rights, including access and benefit sharing with regard to biodiversity.
- Promotion of science for sustainability – in combination with local knowledge systems – and pro-poor technologies.
- Application of Rio principles, including 'the precautionary approach', 'common but differentiated responsibilities', and 'the polluter pays principle' (Rio Declaration on Environment and Development, 1992).
- Access to and sharing of information, including access to private sector information.
- Empowerment of local people, with specific attention to women and youth, indigenous groups, and minorities.
- Ensuring access to health services and cheap and safe medicines.
- Ensuring women's and men's sexual and reproductive rights, including their access to reproductive health services.
- Prevention and resolution of international, regional and local conflicts and wars that jeopardize natural resources and are often fuelled by conflicts over such resources.

All these measures need a gender-specific perspective and approach.

It is necessary not only to focus on one of the levels mentioned above (A, B, C and D), but to ensure that interventions take place and improvements are made at all levels – in a combined effort. Such an integrated approach does not necessarily mean that one organization or institution should tackle all these levels, but that there is coverage of all these aspects by different actors and that their activities are well-coordinated. An important prerequisite is therefore that in each of these areas there is enough awareness, understanding and recognition of both environmental and social aspects – including gender – as well as macroeconomic and political aspects.

This means, for example, that while undertaking activities at the eco-sphere level, the actor should also look into the social and gender aspects in order to prevent negative backlashes on gender equality and to ensure contribution to its enhancement. It is important to be aware that what is good for the environment is not automatically good for specific social groups, such as women. And that what is good for women does not

Box 2.6 *Women's roles in climate change mitigation and adaptation*

Climate change policies and actions are necessary to mitigate climatic changes, and to adapt to those changes that are already occurring and cannot be prevented. Important lessons can be learned from disaster risk reduction, and management. A crucial lesson is that women's active involvement in the development and implementation are important prerequisites to ensure that mitigation and adaptation policies, mechanisms and practices benefit local men as well as women. Also, in the design and promotion of mitigating strategies, women can play a crucial role, for example, in promoting more sustainable lifestyles, education and raising awareness. Therefore women have organized globally to have their voices be heard in international negotiations, and at national and local levels (see Chapters 7 and 8).

automatically improve the environment (NEDA, 1997). Activities, for example, in the area of sustainable land-use, do not necessarily improve women's lives in the short term; these could even add to their work-burden as the work could be more labour-intensive. On the other hand, a more sustainable system could significantly improve women's and men's livelihoods. An evaluation in Bangladesh, for example, showed that women themselves appreciated sustainable land-use activities, as these increased their access to and control over resources and ways of production, as well as their social status (DGIS and Novib, 1994). Similarly, nature conservation measures can exclude local communities from access to important resources (UNEP, 2005), whereas women can also be important drivers of and actors in nature conservation activities. This underlines the need for active involvement and participation of local communities, especially women, in environment activities. In order to promote such an approach, the International Union for Conservation of Nature (IUCN) has developed a series of participatory methodologies to enhance equity in conservation and environmental management (Aguilar et al, 2002).

At organizational level, several measures are needed to enhance an integrated approach and to promote gender mainstreaming in environment:

- Basic understanding of the relationship between gender, environment and development, as well as of the relevance of gender-specific approaches in environmental work.
- Capacity-building, including training, on gender and environment.
- Use of gender-specific approaches, tools and instruments, such as gender-environment analysis, participatory methods, gender-specific

environmental data, criteria and indicators, gender budgeting and gender-specific monitoring and evaluation.

- Cooperation and coordination between organizations and departments working in the field of environment, sustainable development, gender equality and women's empowerment. In these efforts grassroots women, NGOs, government agencies, and academia should also be involved.
- Development and support of activities in the field of gender and environment, such as projects on women's empowerment in sustainable energy technologies, promotion of women's access to and control over land and other natural resources, capacity-building of local organizations and development of livelihood alternatives for women and their families living in seriously degraded or threatened environments.

Conclusions

This chapter has emphasized that the interactions between humans and their environment are intense, complex and gender-differentiated. Local women and men interact in diverse ways with their environment, and the environmental functions and ecological services are often different for women and men. Environmental change has gender-differentiated impacts and might even increase gender inequality and burdens on women. Although gender equality is not a prerequisite for environmental sustainability *per se*, it significantly contributes to environmental conservation and management, as women's rich contributions and potentials in this area can manifest themselves optimally. On the other hand, environmental sustainability generally makes women's lives easier and opens opportunities for full development. However, it is not automatically a guarantee for gender equality. Other forms of social differentiation such as class, ethnicity, religion and age, are important determinants as well, and also need specific attention, alongside gender issues.

Sustainable development requires a focus on both environmental and social aspects of societies and the interlinkages between both domains. The structural relationship between women and men in society shapes the functions that the environment and natural resources have for both genders, as well as the role that women and men play in environmental use and management. These dynamics are visible in the realities in which rural and urban communities live. This leads to the realization that efforts towards sustainable development and environmental conservation need a gender-specific and participatory approach. On the other hand, it also makes clear that, in order to enhance gender equality and to empower local women and men, an understanding of the physical context in which people live is

Box 2.7 *Climate change and gender call for an integrated approach*

In the complex web of relationships between communities and their environment, climate change is a multiplier of environmental changes and has major impacts on the environmental functions and services on which women and men depend and with which they interact. These impacts and consequences are not gender-neutral and often affect women and girls in a more direct or severe way in their roles, responsibilities and opportunities. At the same time, experiences of women and men in coping with and adapting to sudden and gradual environmental stress and climatic changes can be valuable contributions to dealing with these.

The present-day and future challenges of climate change call for an integrated, interdisciplinary approach in which not only technical– environmental aspects are covered, but that also clearly take into consideration the social context and conditions in which these pro- cesses take place. Women and men – of diverse age, class and eth- nicity – have different needs, priorities and possibilities in mitigating and adapting to climatic changes. Therefore climate change policies, mechanisms and actions can be more effective and enhance equality if they take into account gender aspects in their development, planning and implementation, and lead to empowerment of local communities and local women in particular.

needed. It calls for an integrated, holistic approach that focuses not only on improving the situation in specific domains, but also invests in improving the inter-linkages between those domains. As has been indicated by Capra (1975) it is actually the quality of the relationships that determines the quality of a complex system. Therefore, not only multidisciplinary but also interdisciplinary approaches are needed in order to promote sustainable development at a local level, where people actually live.

References

Agarwal, B. (1998) 'The gender and environment debate', in R. Keil (ed), *Political Ecology: Global and Local*, Routledge, London and New York

Agarwal, B. (2000) 'Conceptualising environmental collective action: Why gender matters', *Cambridge Journal of Economics*, vol 24, pp283–310

Aguilar, L. and Castañeda, I. (2001) *About Fishermen, Fisherwomen, Oceans and Tides: A Gender Perspective In Marine-coastal Zones*, International Union for Conservation of Nature (IUCN), San José, Costa Rica

Aguilar, L., Castañeda, I. and Salazar, H. (2002) *In Search of the Lost Gender: Equity in Protected Areas*, International Union for Conservation of Nature (IUCN), San José, Costa Rica

Arnold, J. E. M., Koehlin, G., Persson, R. and Shepherd, G. (2003) 'Fuelwood Revisited: What has changed since the last decade?', in *Occasional Paper no 39*, Centre for International Forestry Research (CIFOR), Bogor

Arumugam, V. (1992) *Victims Without Voice: A Study of Women Pesticide Workers in Malaysia*, Tenegatita and Pesticides Action Network Asia-Pacific, Penang

Awumbila, M. and Momsen, J. H. (1995) 'Gender and the environment: Women's time use as a measure of environmental change', *Global Environmental Change*, vol 5, no 4, pp337–346

Braidotti, R. (ed) (1993) *Women, Environment and Sustainable Development: Towards a Theoretical Synthesis*, Zed Books, London

Bulle, S. (1999) *Issues and Results of Community Participation in Urban Environment: Comparative Analysis of Nine Projects on Waste Management*, ENDA/WASTE, UWEP Working Paper 11, Gouda, Netherlands

Capra, F. (1975) *The Tao of Physics*, Bantam Books, New York

Cecelski, E. (1985) *The Rural Energy Crisis, Women's Work and Basic Needs: Perspectives and Approaches to Action*, International Labour Organization (ILO), Geneva

Chant, S. and Pedwell, C. (2008) *Women, Gender and the Informal Economy: An Assessment of ILO Research and Suggested Ways Forward*, ILO, Geneva

Chen, B. H., Hong, C. J., Pandey, M. R. and Smith, K. R. (1990) 'Indoor air pollution in developing countries', *World Health Statistics Quarterly*, vol 43, no3, pp127–138

Chen, M., Vanek, J., Lund, F. and Heintz, J. (2005) *Progress of the World's Women 2005*, United Nations Development Fund for Women (UNIFEM), New York

Choo, P. S., Nowak, B. S., Kusakabe, K. and Williams, M. J. (2008) 'Guest Editorial: Gender and Fisheries', *Development*, vol 51, pp176–179

CIFOR (Centre for International Forestry Research) (2003) *Annual Report*, CIFOR, Bogor

Dahlberg, F. (ed) (1983) *Woman the Gatherer*, Yale University Press, New Haven and London

Dankelman, I. (2007) 'Gender and environment: Lessons to learn', in E. Barbieri-Masini (ed) 'Human Resource System Challenges', in UNESCO, *Encyclopedia of Life Support Systems*, EOLSS Publishers; developed under auspices of UNESCO, Oxford, www.eolss.net/outlinecomponents/Human-Resource-System-Challenges.aspx, accessed 20 March 2010

Dankelman, I. and Davidson, J. (1988) *Women and Environment in the Third World: Alliance for the Future*, Earthscan, London

David, R. (1995) *Changing Places? Women, Resource Management and Migration in the Sahel*, SOS Sahel, London

Davidson, J., Myers, D. and Chakraborty, M. (1992) *No Time to Waste: Poverty and Global Environment*, Oxfam GB, Oxford

d'Eaubonne, F. (1974) *Le Feminisme ou la Mort*, Pierre Horay, Paris

deVries, I. C. and Keuzenkamp, S. (2000) *Compendium Women and Habitat*, University of Nijmegen, Department of Spatial Planning, Nijmegen

DGIS and Novib (1994) *NGOs and Sustainable Land Use in Bangladesh: A study of the programmes regarding sustainable land use of three partner organizations of Novib in Bangladesh*, DGIS and Novib, The Hague

Ertürk,Y. (2009) 'Statement on the occasion of International Women's Day', Geneva, 6 March 2009, www.un.org/womenwatch/feature/financialcrisis, accessed 3 February 2010

FAO (Food and Agriculture Organization of the United Nations) (1997) 'Brahui Women's Indigenous Knowledge of Medicinal Plants', in *Inter-regional Project for Participatory Upland Conservation and Development,* Working Paper 5, FAO, Quetta

FAO (2007) *Forests and Energy: Key Issues,* FAO, Rome

FAO (2009) 'Facts and figures on gender and food security – division of labour', factsheet, FAO, Rome

Forsyth,T. (2005) *The Encyclopedia of International Development,* Routledge, London and New York

Friedman, L. (2009) 'Coming Soon: Mass Migrations Spurred by Climate Change', *New York Times,* 2 March 2009, www.nytimes.com/cwire/2009/03/02/02climatewire-facing-the-specter-of-the-globes-biggest-and--9919.html?pagewanted=1

Goldman, M. (1997) *Privatizing Nature: Political Struggles for the Global Commons,* Pluto Press, London

GWA (Gender and Water Alliance) (2003) *The Gender and Water Development Report: Gender Perspectives in the Water Sector,* GWA/International Water and Sanitation Centre (IRC), Delft, Netherlands

Hardin, G. (1968) 'Tragedy of the Commons', *Science,* vol 162, no 3859, pp1243–1248

Howard, A. (1940) *An Agricultural Testament,* Oxford University Press, Oxford

Howard, P. (ed) (2003) *Women and Plants: Gender Relations in Biodiversity Management and Conservation,* Zed Books, London

Hunter, L. M. and David, E. (2009) *Climate Change and Migration: Considering the Gender Dimensions,* Institute of Behavioral Science (IBS), University of Colorado at Boulder, Population Program, POP2009-13, report submitted to UNESCO, volume on Migration and Climate

Hyndman, J. (2005) 'Feminist geopolitics and September 11', in L. Nelson and J. Seager (eds) *A Companion of Feminist Geography,* Blackwell Companions on Geography, Blackwell Publishing, Malden, Oxford, Carlton, pp565–577

ILO (International Labour Organization) (1967) *Report form for the Maximum Weight Convention 1967 (no 127),* ILO, Geneva

ILO (2009a) *Global Employment Trends for Women Report 2009,* ILO, Geneva

ILO (2009b) *Global Employment Trends Update,* May 2009, ILO, Geneva

Imran, S. (2002) 'Empowering local communities for natural resources management in Baluchistan, Pakistan', in S. Cummings, H. van Dam and M. Valk (eds), *Natural Resources Management and Gender: A Global Source Book,* KIT Publishers/Oxfam GB, Amsterdam/Oxford

IPCC WG2 (Working Group II) (2007) *Climate Change 2007: Impacts, Adaptation and Vulnerability: Contribution of Working Group II to the Fourth Assessment Report of the Intergovernmental Panel on Climate Change,* M. L. Parry, O. F. Canziani, J. P. Palutikof, P. J. van der Linden and C. E. Hanson (eds), Cambridge University Press, Cambridge

Jansen, W. (2004) 'The economy of religious merit: Gender and *ajr* in Algeria', *Journal of North African Studies,* vol 9, no 4, pp1–17

Jansen, W. (2007) 'Economics: Traditional professions: Arab States', in S. Joseph (ed) *Encyclopedia of Women and Islamic Cultures,* vol 4, E. J. Brill, Leiden, pp262–266

Johnson, V., Hill, J. and Ivan-Smith, E. (1995) *Listening to Smaller Voices: Children in an Environment of Change*, Actionaid, Somerset

Jordans, E. H. and Zwarteveen, M. Z. (1997) *A Well of One's Own: Gender analysis of an irrigated program in Bangladesh*, IMII Country Paper no 1, IIMI, Colombo

Kelkar, G. and Tshering, P. (2004) *Themes for Celebrating Women*, International Centre for Integrated Mountain Development (ICIMOD), Kathmandu

Leach, M. (1992) 'Gender and the environment: Traps and opportunities', *Development in Practice*, vol 2, no 1, pp12–22

Leduc, B. (2009) *Gender and Climate Change in the Himalayas*, background paper for the e-discussion 'Climate Change in the Himalayas: The Gender Perspective', 5–25 October, organized by ICIMOD and Asia Pacific Mountain Network (APMN), ICIMOD, Kathmandu

Lee, R. and De Vore, I. (eds) (1968) *Man the Hunter*, Aldini, Chicago

Lewis, N., Huyer, S., Kettel, B. and Marsden, L. (1994) *Safe Womanhood*, discussion paper, International Federation of Institutes of Advanced Study, Toronto

Mazumdar, V. (1994) *Embracing the Earth: An Agenda for Partnership with Peasant Women*, FAO, New Delhi

McPeak, J. (2002) *Fuelwood Gathering and Use in Northern Kenya*, paper prepared for Meeting of Association for Politics and Life Sciences, Montreal, 11–14 August 2002

Mead, M. (1949 and 2001) *Male and Female*, HarperCollins, New York

Menon, G. (1991) 'Ecological transitions and the changing context of women's work in tribal India', *Purusartha* pp291–314

Merchant, C. (1980) *The Death of Nature: Women, Ecology and the Scientific Revolution*, HarperCollins, New York

Meuffels, S. (2006) 'Sustainability of Different Farming Systems in Garhwal Himalaya, India', Biological Farming Systems, Wageningen University and Research Centre, Wageningen

Millennium Ecosystem Assessment (2003) *Ecosystems and Human Well-Being: A Framework for Assessment*, Island Press, Washington, DC

Muller, M. and Schienberg, A. (1998) *Gender and Urban Waste Management*, Global Development Research Center, Kobe

Murdock, G. P. and White, D. C. (1969) 'Standard Cross-Cultural Sample', *Ethnology*, vol 8, no 4, pp329–369

NEDA (Netherlands Development Assistance) (1997) 'Gender and Environment: A Delicate Balance Between Profit and Loss', in *Women and Development Working Paper*, Ministry of Foreign Affairs, The Hague

Nozawa, C. M. C. (1998) 'Empowerment of women in Asian fisheries', in M. J. Williams (ed) *Women in Asian Fisheries: International Symposium on Women in Asian Fisheries*, Chiang Mai, 13 November 1998, WorldFish Center, Penang

Oracion, E. G. (2001) 'Filipino women in coastal resources management: The need for social recognition', in *SMA Working Papers Series*, Silliman University, Dumaguete City

Ortner, S. (1974) 'Is female to male as nature is to culture?' in M. Roslado and L. Lamphere (eds) *Women, Culture and Society*, Stanford University Press, Stanford

Owen, L. (1998) 'Frauen in Altsteinzeit: Mütter, Sammlerinnen, Jägerinnen, Fisherinnen, Köcherinnen, Herstellerinnen, Künstlerinnen, Heilerinnen' in Auffermann, B. and Weniger, G. C. (eds), *Frauen, Zeiten and Spuren*, Neanderthaler-Museum, Mettmann

Parbring, B. (2009) 'Gender Impact of Climate Change', report of the debate 'The Heat is On: Debate on Gender and Climate Change in North and South', Nordic Council of Ministers, 4 December 2009, www.norden.org/en/news-and-events/news/gender-impact -of-climate-change, accessed 3 February 2010

Pijnappels, M. (2006) *Gender and Natural Disasters: Women's Role and Position*, Centre for Sustainable Management of Resources (CSMR), Radboud University, Nijmegen

Radkau, J. (2008) *Nature and Power: A Global History of the Environment*, Cambridge University Press, Cambridge

Rengam, S. V., Bhar, R. H., Mourin, J. and Ramachandran, R. (2007) *Resisting Poisons, Reclaiming Lives! Impact of Pesticides on Women's Health*, Pesticide Action Network Asia and the Pacific (Pan-AP), Penang

Rio Declaration on Environment and Development (1992), www.unep.org/Docu ments.Multilingual/Default.asp?DocumentID=78&ArticleID=1163, accessed 3 February 2010

Rocheleau, D. E. (1995) 'Gender and Biodiversity: a feminist political ecology perspective', in S. Joekes, M. Leach and C. Green (eds), *Gender Relations and Environmental Change*, Institute for Development Studies (IDS), Brighton

Rocheleau, D. E., Thomas-Slayter, B. and Wangari, E. (1996) *Feminist Political Ecology: Global Issues and Local Experiences*, Routledge, London and New York

Seager, J. (1993) *Earth Follies: Feminism, Politics and the Environment*, Zed Books, London

Saito, K. A., Mekonnen, H. and Spurling, D. (1994) 'Average time spent in agricultural activities by gender in Burkina Faso, Kenya, Nigeria, and Zambia', in C. M. Blackden and Q. Wodon (eds) (2006) *Gender, Time Use, and Poverty in Sub-Saharan Africa*, World Bank Working Paper no 73, World Bank, Washington, DC, p18

Shiva, V. and Dankelman, I. (1992) 'Women and Biological Diversity', in A. O. D. Cooper (ed) *Growing Diversity, Genetic Resources and Local Food Security*, Intermediate Technology Publications, London

Shiva, V. (1988) *Staying Alive: Women, Ecology and Development*, Zed Books, London

Shiva, V., Dankelman, I., Singh, V. and Negi, B. (1990) *Biodiversity, Gender and Technology in Agriculture, Forestry and Animal Husbandry*, University of Wageningen, Dehra Dun and Wageningen

Singh, V. (1987) 'Hills of Hardship', *The Hindustan Times*, 18 January 1987

Singh, V. (1988) *Development of Sustainable Development in Garhwal*, G. B. Pant University, Ranichauri

Slater, R. and Twyman, C. (2003) *Hidden Livelihoods? Natural Resource-dependent Livelihoods and Urban Development Policy*, Working Paper no 225, Overseas Development Institute (ODI), London

Stanley, A. (1981) 'Daughters of Isis, daughters of Demeter: When women sowed and reaped', *Women's Studies International Quarterly*, vol 4, no 3, pp289–304

Steady, F. C. (ed) (1993) *Women and Children First: Environment, Poverty and Sustainable Development*, Schenkman Books, Rochester

Traoré, D., Keita, M., Sacko, B. and Muller, M. (2003) *Citizen Involvement in Clean-Up Activities in Bamako: Lessons from an Action Research Project in Commune IV*, UWEP Working Document 13, CEK & Waste, Gouda

UBINIG (Unnayan Bikalper Nitnirdharoni Gobeshona) (ed) (2003) *'Beesh': Poisoning of women's lives in Bangladesh*, UBINIG (Policy Research for Development Alternatives), Dhaka in collaboration with PAN-AP

UN (2009) *The Millennium Development Goals Report 2009*, UN, New York

UN Development Project (2005) *Investing in Development: A Practical Plan to Achieve the Millennium Development Goals*, Millennium Project, New York

UNDP (United Nations Development Programme) (2002) *Mainstreaming Gender in Water Management: A Practical Journey to Sustainability: A Resource Guide*, UNDP, New York

UNEP (United Nations Environment Programme) (2005) *Final Report Online Discussion on Women and the Environment*, UNEP, Nairobi

UNFPA (United Nations Fund for Population Affairs) (2002) *State of the World Population 2002*, UNFPA, New York

UNFPA (2006) *State of the World Population 2006: A Passage to Hope: Women and International Migration*, UNFPA, New York

UNICEF/WHO (United Nations Children's Fund/World Health Organization) (2008) *Progress on Drinking Water and Sanitation: special focus on sanitation*, UNICEF and WHO, New York/ Geneva

UN-WAPP (2006) *The United Nations World Water Development Report 2*, UNESCO and Berghahn Books, Paris and London

UN-Water (2006) *Coping with Water Scarcity*, FAO, Rome

WEDO (Women's Environment & Development Organization) (2003) *Diverting the Flow: A Resource Guide to Gender, Rights and Water Privatization*, WEDO, New York

WEDO/UNIFEM (United Nations Development Fund for Women) (2010) *Gender and Climate Change Assessment: Caribbean region*, WEDO, New York, unpublished report

WHO (World Health Organization) (2004) *Gender, Health and Work*, WHO, Geneva

WHO (2006) *Fuel for Life: Household Energy and Health*, WHO, Geneva

Wickramasinghe, A. (1994) *Deforestation, Women and Forestry*, INDRA International Books, Amsterdam

WIEGO (Women in Informal Employment Globalizing and Organizing) (2009) *Informal Workers in Focus: Street Vendors, Waste Collectors, Home-based Workers*, WIEGO, Cambridge, MA

Williams, M. J., Nandeesha, M. C. and Choo, P. S. (2006) *Changing Traditions: A Summary Report on the First Global Look at the Gender Dimensions of Fisheries*, paper read at the Global Symposium on Gender and Fisheries, 7th Asian Fisheries Forum, Penang, 1–2 December 2004

World Bank (2006) 'Gender, Time Use and Poverty in Sub-Saharan Africa', in C. M. Blackden and Q. Wodon (eds) *Gender, Time Use, and Poverty in Sub-Saharan Africa*, World Bank Working Paper no 73, World Bank, Washington, DC

World Bank (2008) *Social Dimensions of Climate Change Report*, World Bank, Washington, DC

Zukang, S. (2009) *Statement to the 53rd session of the Commission on the Status of Women*, New York, 2 March 2009, www.un.org/womenwatch/feature/financialcrisis, accessed 3 February 2010

3

Climate Change, Human Security and Gender[1]

Irene Dankelman

This chapter develops an analysis of the relationship between climate change, human security and gender issues. It describes how climate change and human security are inter-related. It will then focus on gender implications, examining the gender aspects of (natural) disasters and the impacts of climate change on human security, particularly on women's vulnerability. Furthermore, this chapter explores the role that many women play in strengthening human security when climatic changes occur. Based on these explorations, an analytical framework on gender, climate change and human security is presented and discussed at the end of this chapter.

Climate change and human security

Climatic changes result in a variety of direct problems, including increased frequency of extreme weather events, flooding, storms, drought, desertification, increases in sea temperatures, cold and heatwaves, the melting of glaciers and permafrost. In the long run, the rise in sea levels and abrupt changes in currents pose major threats to coastal areas, ecosystems and geophysical cycles. These developments have significant ecological, social, economic and political impacts, including effects on biodiversity, food production, water availability, intensification of wildfires, mud-streams, bleaching of corals and changes in epidemic vectors.

The 2007–2008 Human Development Report: Fighting Climate Change: Human Solidarity in a Divided World (UNDP, 2007a) concludes that climate change threatens progress towards development itself and also progress towards meeting the 2000 UN Millennium Development Goals (MDGs) in particular. According to the report, climatic changes will undermine the raising of the Human Development Index in many countries (UNDP, 2007a). This perspective moves climate change away from a purely technical subject and brings it to the centre of (sustainable) development policies and strategies.

As Kok and Jäger (2009, p13) conclude, the rate of global environmental change that we are currently witnessing has an increasing impact on human well-being, as the provision of ecosystem services, such as food production, clean air and water or a stable climate, are severely and increasingly threatened.

The nature and extent of climate change not only hinder human development and environmental conservation, but also form a major threat to human security at livelihood levels. In a statement at the Royal United Services Institute's January 2007 Conference on Climate Change: The Global Security Impact, John Ashton, the UK Foreign Secretary's Special Representative for Climate Change said: 'There is every reason to believe that as the 21st century unfolds, the security story will be bound together by climate change... Climate change is a security issue because if we don't deal with it, people will die and states will fail' (in Littlecott, 2007, p1).

On 17 April 2007, for the first time in its history, the UN Security Council took up the issue of climate change as an important human security challenge and agreed that climate change is threatening geo-political security. At that meeting, the former UK Foreign Secretary Margaret Beckett warned that drought and crop failure could cause intensified competition for food, water and energy, saying 'it is about our collective security in a fragile and increasingly interdependent world' (UN, 2007).

Due to flooding, disease and famine, climate change may spark conflict between and within countries, as resources and safe places become scarcer and disasters destroy livelihoods, increasing the number of migrants and refugees. Edward et al (2004) explain, for example, how shortfalls in seasonal rains that result in drought and economic distress increase the likelihood of civil war by up to 50 per cent.

On the other hand, scholars such as Betsy Hartmann, warn against the dangers and obscurity of linking climate change with national security.

> *While environmental changes due to global warming could exacerbate already existing economic and political tensions in many locations, the threat scenarios being banded about in security circles are wildly speculative and based on racialized stereotypes of poor people. They ignore the ways many poorly resourced communities manage their affairs without recourse to violence.* (Hartmann, 2009, p8).

Such a warning stresses the need to avoid making climate change a pawn of political interests and to focus on human security – the security of survival, of livelihood and dignity – instead.

Climate change also has major economic implications. According to the 2006 report of Sir Nicholas Stern et al, the costs of adapting to climate change could be as much as 10 per cent of the global economic output (Stern et al, 2006). But climate change is not just a political and economic challenge. Most of all, it is an ecological and humanitarian issue, where the livelihoods of numerous communities are threatened and their security is at stake. Wisner et al (2007) in their analysis of the interactions between climate change and human security issues show that human security is threatened and conflicts arise where climate change impacts on food, health and water availability.

Authors also see potential conflicts occurring between food and fuel production in competition for land and other resources and in the promotion of biofuels and other climate change mitigation measures. Even adaptation can contribute to inequity and become a source of conflict. O'Brien (2007) warns that, in our globalizing world, human security in one place or for one group is increasingly linked to the actions and outcomes of other groups in other places. Climate change mitigation and adaptation potentially create new inequities, vulnerabilities and insecurities.

The UNEP GEO-3 report (2002) defines vulnerability as: 'The interface between exposure to physical threats to human well-being and the capacity of people and communities to cope with those threats'. Wisner et al (2004, pp4 and 11) define vulnerability as 'the characteristics of a person or group and their situation influencing their capacity to anticipate, cope with, resist and recover from the impact of natural hazard'. As physical circumstances differ from place to place, and people and communities are very diverse, vulnerabilities and capacities are usually place-based and context-specific (Kok and Jäger, 2009). The consequences of climate change are closely related to the context in which individuals or groups experience the changes (O'Brien, 2007). Although climate change affects everyone regardless of class, race, age and gender, its impacts are more heavily felt by poor persons, communities and nations. Often climate change tends to magnify existing inequalities, and increases chronic instability. The Intergovernmental Panel on Climate Change (IPCC) concluded: 'Poor communities can be especially vulnerable, in particular those concentrated in high-risk areas. They tend to have more limited adaptive capacities, and are more dependent on climate-sensitive resources such as local water and food supplies' (IPCC, 2007, p9).

Based on these observations, many authors (Wisner et al, 2004; Lambrou and Piana, 2005; Oswald Spring, 2007) argue that a vulnerability approach to natural disasters, including climate change, is necessary, as 'the risks involved in disasters must be connected with the vulnerability created for many people through their normal existence'. This means that

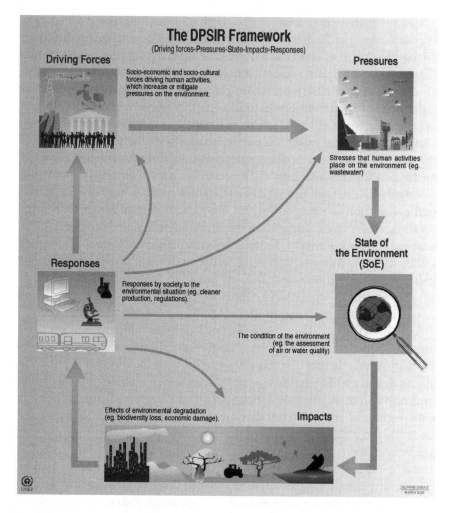

Figure 3.1 *Drivers-pressures-state-impacts-response (DPSIR) framework*

Source: Global International Water Assessment (2001), European Environment Agency, Copenhagen. Retrieved from UNEP/GRID-Arendal Maps and Graphics Library, http://maps.grida.no/go/graphic/ dpsir_framework_for_state_of_environment_reporting, accessed 19 February 2010. Design: Delphine Digout, UNEP/GRID-Arendal.

there is a need 'to identify, delineate, and understand those driving forces that increase or decrease vulnerability at all scales'(Cutter, 2003, p7).

In this respect the drivers–pressures–state–impacts–response (DPSIR) framework is a useful model (see Figure 3.1). In the figure the **drivers** of change are socio-economic and socio-cultural forces driving human activities, which increase or mitigate pressures on the environment. **Pressures** reflect the stressors that human activities place on the environment; and the **state** is the condition of the environment. **Impacts** are the effects of environmental changes on human society, and **responses** indicate the responses that the society applies to change the environmental situation.

Apart from seeing climate change as an apocalyptic global problem that has to be tackled with all strength at all levels, some authors point to the fundamental challenges and opportunities that climate change brings about. 'Ironically, climate change offers humanity an opportunity for a quantum leap in sustainable development and peace making'(Wisner et al, 2007, p11). O'Brien (2007) regards climate change as an opportunity in history to address inequities and enhance human security, and as an extraordinary opportunity for responding to and creating social change. As Margaret Beckett emphasized during the special session of the Security Council in April 2007: 'Climate change can bring us together, if we have the wisdom to prevent it from driving us apart'(UN, 2007).

Gender aspects of natural disasters

In many societies, vulnerability to natural disasters differs for women and men. In many cases, but not always, women are more vulnerable to disasters than men through their socially constructed roles and responsibilities, and because they lack adequate power and assets (Pan American Health Organization, 1998; Mitchell et al, 2007). Important lessons about the gender-specific implications of disasters can be learned from existing gender-disaster literature; from this body of knowledge many consequences of climatic changes on women and men can be extrapolated.[2]

In general, women have less access to resources that are essential in disaster preparedness, mitigation and rehabilitation. Gendered divisions of labour often result in the over-representation of women in agricultural and informal sectors, which are vulnerable to disasters. Lack of energy sources, clean water, safe sanitation and health challenges, often put extra burdens on women's shoulders, adding to their reproductive and care-giving tasks (Enarson, 2000). Therefore when a slow or sudden natural disaster hits, this adds to the double burden of productive and reproductive labour (Patt et al, 2007). In the face of disaster, men are often the first to migrate, while in many societies, socio-cultural norms and care giving responsibilities prevent women from migrating to look for shelter and work when a disaster hits. They are likely to be hindered by the fact that they are less mobile, more confined to the house and have less decision-making power. All of which contributes to their lack of participation and their limited access to information regarding potential hazards, disaster-preparedness and possible adaptation strategies. These realities deprive women of opportunities to look for alternative sources of income and livelihood, adversely affecting their coping capacities.

The direct effects of natural disasters on women and men is diverse. In a study by the London School of Economics, the University of Essex and the Max-Planck Institute of Economics, a sample was analysed of 141 countries in which natural disasters occurred during the period 1981–2002 (Neumayer and Plümper, 2007). The main findings of this study were: (a) natural disasters lower the life expectancy of women more than that of men; (b) the stronger the disaster, the stronger this effect on the gender gap in life expectancy (generally life expectancy is higher for women than it is for men; if the gender gap in life expectancy decreases due to an event it means that relatively more women die, or they die at an earlier age); (c) the higher the socio-economic status of women, the weaker this effect on the gender gap in life expectancy. The conclusion is that it is the socially constructed gender-specific vulnerability of women and men that leads to the relatively higher female disaster mortality rates compared to those of men.

Gender-differentiated roles don't always result in higher losses for women. For example, immediate mortality caused by Hurricane Mitch in Central America was higher for men, not only because they were engaged in outdoor activities when the disaster struck, but also because they tended to be more over-confident in their behaviour toward risk (Bradshaw, 2004). Empirical studies reveal that women and men act and make decisions differently. Whereas men are more risk-taking, women tend to be more risk-averse. The studies show that in general men are more over-confident, thinking that they can predict and handle the future themselves, whereas women are more willing to adapt their strategies and behaviour. Women usually listen to external advice, whereas men will not easily ask for directions. In general, women are more aware of social bonds, showing greater reciprocity and altruism. However, when social bonds are weak, men can be observed to be more cooperative than women. These findings have major implications for disaster management and could form important underlying motives for women's and men's reactions to hazards (Brown-Kruse and Hummels, 1993; Patt et al, 2007).

The often disadvantaged position of women means greater difficulty in coping with disasters. For example, in a country such as Bangladesh, where women are more calorie-deficient than men, women generally have more problems recovering from the negative effects that flooding has on their health (Cannon, 2002). An increase in the number of female-headed households – caused by male out-migration – also amplifies women's responsibilities and vulnerabilities during natural disasters. After a disaster hits, there are often inadequate assets and facilities available for women to cope with the consequences for their households and themselves. In general, disaster relief efforts pay insufficient attention to women's reproductive and sexual health after a disaster and, as a result, their health suffers disproportionately.

Often in the aftermath of disasters, an increase of domestic and sexual violence occurs. During and after disasters such as long periods of drought, more girls drop out of school to reduce household expenses by saving on school fees, or to assist in the household with tasks such as fetching water, or as a result of pregnancy and early marriage (Eldridge, 2002). A 2001 study in Malawi showed that female children are married off early in times of drought, usually to older men with numerous sexual partners. They were even forced to sell sex for gifts or money, which resulted in the accelerated spread of HIV/AIDS in the country (Malawi Government, 2001).

Lower levels of education reduce the ability of women and girls to access information – including early warning mechanisms – and resources, or to make their voices heard. This is an extra challenge when women want to make innovative changes in their livelihoods.

Gender, security and climate change

Although climate change affects everyone, its drivers and effects are not gender neutral. Climate change magnifies existing inequalities, reinforcing the disparity between women and men in their vulnerability to and capability to cope with changing climatic and environmental conditions (UNDP, 2007a; Mitchell et al, 2007).

Worldwide, as the majority of the world's poor and disadvantaged, women are the most vulnerable to the effects of climate change and particularly their security is at stake (WEDO, 2007). As was indicated above, poor women are more likely to become direct victims – through mortalities and injuries – of natural disasters, and the same is true for major climate change events such as hurricanes and flooding (Neumayer and Plümper, 2007). During such natural disasters, often more women die than men because they are not warned, cannot swim or do not leave the house alone (UNFCCC, 2005).

Women made up 55–70 per cent of the Banda Aceh, Indonesia, tsunami deaths in 2004, and in the worst affected village – Kuala Cangkoy in the North Aceh district – 80 per cent of the deaths were women (UNIFEM, 2005; Oxfam, 2005). According to BBC News Online, of the 2003 French heatwave death toll of 15,000, about 70 per cent were women. These numbers, however, should be approached with caution: women also make up the majority of the elderly and that age group is most affected by high temperatures.

When poor women lose their livelihoods, they slip deeper into poverty and the inequality and marginalization they suffer from increases. In 2005 Hurricane Katrina in the US entrenched poor African-American women, who were already the most impoverished group in the nation, into deeper levels of poverty (WEDO, 2007). In that case also ethnicity and class were important determinants of the effects of the disaster.

In many societies worldwide, often more women than men are affected in their multiple roles as food producers and providers, as guardians of health, caregivers to the family and community and as economic actors. As access to basic assets and natural resources, such as shelter, food, fertile land, water and fuel, becomes hampered, women's workload particularly increases. As poor families, many of which are headed by females, often live in precarious situations, on lowlands, along dangerous riverbanks or on steep slopes, they are in danger of loosing their shelters. People's environmental, food, water, energy, shelter and economic security are all negatively influenced, with major implications for local women and men, and their children.

Lack of natural resources, caused by flooding, drought and erratic rainfall particularly cause women to work harder to secure natural resources and livelihoods. In such situations, women and girls have less time to receive education or training, earn an income or to participate in governing bodies. As a UNDP report of 2007 stated:

> *Loss of livelihood assets, displacement and migration may lead to reduced access to education opportunities, thus hampering the realization of Millennium Development Goal 2 (MDG2) on universal primary education. Depletion of natural resources and decreasing agricultural productivity may place additional burdens on women's health and reduce time for decision-making processes and income-generating activities, worsening gender equality and women's empowerment (MDG3).* (UNDP, 2007b, p1)

Relocation of people has severe impacts on social support networks and family ties – mechanisms that have a crucial value, particularly for women's coping capacities (Patt et al, 2007). Conflict that arises from a shortage of natural resources or of safe places to go, amplifies existing gender inequalities, and disaster literature learns that in many of those situations, violence against women increases.

Coping strategies: Efforts to strengthen security

Too often women are seen primarily as the main victims of climate change and not as positive agents of change and contributors to livelihood adaptation strategies. As highlighted by Enarson (2000) and O'Brien (2007) natural disasters could also be a unique opportunity to challenge and change gender inequality in society. Women have been willing and able to take an active role in what are traditionally considered 'male' tasks in responding to disasters. They have proved effective in mobilizing the community to respond to disasters, and in disaster preparedness and mitigation. For example, after Hurricane Mitch struck in 1998 in Central America, women in several communities engaged in disaster-proof housing construction (Schrader and Delaney, 2000).

The disaster was also considered as an opportunity to address gender inequality. The non-governmental organization (NGO) Puntos de Encuentro in Nicaragua organized the information campaign 'Violence against women is one disaster that men can prevent'. The campaign proved effective in changing men's attitudes towards violence against women, and therefore tackled existing power structures (www.puntos.org.in, Pan-American Health Organization, 1998).

Usually women have fewer assets than men to recover from natural disasters, including climatic changes, and in many countries they do not own land that can be sold to secure income in an emergency. Among the problems women identify when having to adapt to climate change are lack of safe land and shelter, lack of other assets and resources, limited access to material and financial resources, lack of relevant skills and knowledge, high prices of agricultural inputs and other materials, and cultural and other barriers limiting their access to services (Mitchell et al, 2007).

However, local women and men worldwide are starting to cope with a changing climate and beginning to articulate their priorities and needs for securing and sustaining their livelihoods. There can be major differences between coping with the existing situation – making the best of it with the limited resources available – and adapting, often a more sustainable form that helps to safeguard and regain a livelihood in a changed or changing local context. Coping strategies are not always sustainable or healthy, such as skipping meals, using inferior energy sources or money lending (Christ, 2008). Local coping and adaptation strategies in which women are involved aim to strengthen the direct security of their families and communities.

Recent experience shows that local coping and adaptation strategies in which women play a crucial role, include:

- Moving to safer places: Moving to higher locations, making of temporary shelters, increasing the plinth level of their houses, and migration.
- Saving of assets: Storing seeds and moving livestock to higher places.
- Adaptation of agricultural practices: Switching to crops and varieties that are flood or drought resistant; multiple cropping and intercropping practices; alternative irrigation facilities; changing cultivation to more easily marketable crop varieties or to other animals (for example, rearing goats and poultry farming in Nepal, or ducks instead of poultry in Bangladesh).
- Energy-saving: Using alternative energy-related technologies (solar, biogas, improved cooking stoves).
- Dietary adaptations: Skipping meals or eating non-traditional foods (such as water hyacinth), or preserving food to be used in the lean times.
- Earning income or saving money: Working as wage (bound) labourers; borrowing money from money lenders often at high interest rates; secretly saving part of their earnings; distress sale of livestock; and, in situations of major distress, stepping into prostitution.
- Alternative health care: Using and promoting traditional medicine and medical practices.
- Organizing and collective action: Setting up of community-based self-help groups and networks and group savings or systems of group labour (for example through the Nepalese system of *parma*) are popular ways that women apply in coping with disasters (Mitchell et al, 2007; Patt et al, 2007).

In several studies women have voiced their priorities in times of disaster:

- Safety: A safe place to live for their families and themselves, including relocation to safer areas, shelters and adaptation *in situ* by the construction of solid houses; the safe storage of their harvest, seeds and livestock, as well as of utensils, stoves and fuelwood.
- Food security: Support in adaptation in agricultural practices, including crop diversification.
- Information: Secure access to relevant information.
- Services: Access to health services (doctors, pharmacists) and agricultural extension.
- Strengthened capacities: Development of their capacities through training and information – including through exposure and exchange visits about adaptation strategies and livelihood alternatives.
- Resources: Access to resources, including climate-related finances, improved access to credits and markets (Skutsch, 2004).
- Environmental security: Ecological restoration.

These articulated needs show that there is ample room for adequate policy interventions to support local women and men in their efforts to cope with and adapt to climatic changes. There could be important synergies between climate policy and gender equality, such as linking carbon sequestration with poverty reduction, clean energy for cooking that reduces indoor air pollution and related diseases, cheap voltaic energy that lights houses and enables education, and ecosystem improvement promoting sustainable development and making women's tasks less burdensome (Christ, 2008).

Also, through their contributions to mitigation, men as well as women worldwide contribute to climate change mitigation measures as a means of risk aversion, contributing to the security of present and future generations. Particularly through their roles as managers of households, in child care and education, and as consumers, women worldwide – constituting more than 50 per cent of the world population – play a crucial role in this respect. However, their decision-making power and participation in defining appropriate mitigation means, mechanisms and processes are still limited (see Chapter 6).

An analytical framework on gender, climate change and human security

Based on experiences worldwide, an analytical framework on gender, climate change and human security has been developed and presented in Table 3.1. In the framework human security is defined as: (a) security of survival; (b) security of livelihood; and (c) dignity. 'Security of survival' manifests itself in levels of mortality or injury and in human health. 'Security of livelihood' encompasses security of food, water, energy, environmental security, security of shelter and economic security. The third category, 'dignity', can be divided in respect of basic human rights, people's capacities and participation. In the framework, the specific effects of climatic changes on these security aspects and its gender implications are identified. The framework also shows what strategies local people have adopted to cope with these effects, and it identifies policy and other options to strengthen human security in these areas.

Conclusion

This chapter presents a gendered analysis of how climate change impacts on human security. It also assesses what scope exists for affected women and men to participate in improved human security in a scenario of a

Table 3.1 Human security, climate change and women

Human security	Security aspects	Climate change impacts	Impact on women	Coping and adaptive strategies by women	Policy opportunities
Security of survival	Mortality/ injury	• Mortality/ injury from different extreme weather events/ disasters	• Overall more women than men die or are injured	• Searching for safe shelter/ improving homes • Disaster risk reduction and preparedness by women's groups	• Gender specific and sensitive disaster risk reduction and preparedness • Early warning systems addressing both women and men
	Health	• Increase in infectious diseases • Physical and mental stress • Loss of medicinal plants/ biodiversity	• Bearing the burden of taking care of the sick/disabled • Increase in mental stress • Lack of access to reproductive health services • Greater risk of HIV/AIDS due to early marriage, forced prostitution, sexual violence	• Increase in tasks for family care • Use of medicinal plants and application of alternative healing methods	• Access to health facilities, especially reproductive health services for women • Monitoring health situation of most vulnerable groups
Security of livelihood	Food security	• Harvests destroyed • Agricultural production changes/drops • Fish stocks decrease	• Bearing the burden of more time, energy and budget requirements for food production and purchase • Stand in line for humanitarian food distribution • Increase in work burden • Increase in calorie-deficiency and hunger	• Adapting agricultural practices/ switching to other crops/ animals • Saving food, seed and animals • Adapting diet	• Agricultural adaptation: mixed cropping, better suited crops/ livestock • Affordable and ecologically sound agricultural inputs • Nutritional extension services • Secure land rights for women • Credit and marketing facilities • Managing fish stocks for local fishing communities
	Water security	• Lack of water • Pollution and water salination • Flooding	• More time/energy needed to provide water for household/ farm • Increase in work burden • Suffer from water-related health problems	• Water saving, including rainwater harvesting • Purchasing water from water-vendors	• Safeguarding affordable and safe drinking water • Efficient irrigation technologies • Safe sanitation facilities • Preserving wetlands

Energy security	• Lack of biomass fuel • Dysfunctioning hydropower • Disruption in electricity supply	• More time and energy needed to collect fuel • Increase in work burden • Inferior energy sources – more indoor pollution	• Switching to other energy sources • Use of energy-saving devices • Reforestation	• Providing fuel sources, especially clean sustainable energy • Providing and training in using energy saving devices • Ecological restoration
Environmental security	• Environmental processes and services jeopardized	• Poorest women living in insecure environments most affected	• Building more sturdy houses • Clean up and regeneration of environment • Forming advocacy groups	• Ecological restoration • Safe shelters
Shelter security	• Housing, infrastructure and services destroyed	• Limited land rights • Not included in land management • Decrease in mobility	• Building more sturdy houses • Seeking shelter • Migration	• Safe shelters and sturdy homes • Land and housing rights for women
Economic security	• Decrease in income generating opportunities	• Women in informal sector most effected • Household expenses increase • Males migrate – more de facto female headed households	• Saving expenses or money for lean time • Selling of assets and services • Alternative income generating activities	• Affordable and reliable credit and financial facilities for women • Providing alternative livelihood options • Ensuring women's access to climate change funding and technologies
Basic human rights	• Triggers violation of basic human rights	• Increase in domestic violence against women • Suffering from conflicts over resources	• Social networks and groups • Organization of women	• Counselling and legal services • Defending women's rights
Dignity				
Capacity	• Lack of opportunities for education and income generation	• Girls drop out of school • Little time for education/ training income generation	• Self-training, support groups and networks	• Ensuring education, particularly of girls during/after disasters • Skills training • Environmental regeneration • Access to information
Participation	• None or limited involvement in decision-making • Lack of information • Lack of time	• Lack of participation in climate change negotiations, planning and activities • Women-specific priorities neglected	• Organization • Advocacy • Participation	• Ensuring women's participation in planning/decision-making/ climate change and mechanisms • Involving men in gender training • Generate and use sex-disaggregated data

Source: Dankelman (2009)

changing climate. As climate change tends to magnify existing inequalities, with gender inequality being one of the most pervasive, it has major impacts on the security particularly of women and, with them, of children. It is now widely acknowledged that overall negative effects of climate change affects women's livelihoods more than those of men because in many communities they depend on natural resources and the environment for most of their activities and for meeting the basic needs of their families (Diagne Gueye, 2008).

In order to understand the implications of climate change on human security, many lessons can be learned from existing literature and experiences from the disaster-gender field. Women often face greater hurdles than men in a disaster. For example, early warning information does not reach women as easy as men, they face privacy issues in shelters, including the lack of safe sanitation, and they are often subjected to greater domestic abuse and violence during and after a disaster.

Gender-specific climate change vulnerability and adaptive capacity are location and context specific. Already in many countries, women experience the impacts of climate change through the increased frequency and intensity of floods, droughts and cyclones. This changing nature also has an impact on their ability to cope and therefore has major consequences on their security and that of their families. Climate change not only affects women's health and well-being differently from that of men, but also impacts negatively on their work burdens, opportunities and capacities through changes in their livelihood security.

Gender differences must be considered not just in terms of differential vulnerabilities, but also as differential adaptive capacities. Women play a key role in increasing human security by protecting, managing and recovering their household and assets during and after climatic disasters and changes. They are strong advocates for preparedness measures at the community level and have knowledge and capacities to contribute towards coping with and adapting to climatic changes. Local women continue to develop innovative strategies to address climate change impacts and demonstrate diverse coping strategies and mechanisms.

There are several factors that affect people's ability to adapt to climate change and enhance human security: access to assets, protection of their economic livelihood, access to services, political participation in decision-making, access to information and enhanced leadership. All these factors are shaped by the politics of gender (Ahlers and Zwarteveen, 2009). And in this period of a changing climate, it is in those areas where opportunities and duties lay for policy-makers and development practitioners to enhance gender equality and human security.

Notes

1 This chapter is partly based on Chapter 2 that the author prepared for the Women's Environment and Development Organization (WEDO) report on Gender, Climate Change and Human Security: Lessons from Bangladesh, Ghana and Senegal, WEDO, New York, May 2008.
2 A useful resource is the website of the Gender and Disaster Network that presents gender-specific lessons from examined natural disasters, www.gdonline.org.

References

Ahlers, R. and Zwarteveen, M. (2009) 'The water question in feminism: Water control and gender inequities in a neo-liberal era', *Gender, Place & Culture*, vol 16, no 4, pp409–426

Bradshaw, S. (2004) *Socio-economic Impacts of Natural Disasters: A Gender Analysis*, UN Economic Commission for Latin America (ECLA), Serie Manuales 32

Brown-Kruse, J. and Hummels, D. (1993) 'Gender effects in laboratory public goods contributions: Do individuals put their money where their mouth is?', *Journal of Economic Behavior & Organization*, vol 22, p255–267

Cannon, T. (2002) 'Gender and climate hazards in Bangladesh', *Gender and Development*, vol 10, no 2, pp45–50

Christ, R. (2008), 'Women, a most vulnerable group', speech at the Roundtable on Climate Change and Human Security, Vienna, 13 March 2008

Cutter, S. L. (2003) 'The vulnerability of science and the science of vulnerability', *Annals of the Association of American Geographers*, vol 93, no 1, pp1–12

Dankelman, I. (2009) 'Bearing the burden', *UNChronicle*, vol 46, nos 3–4, pp50–53

Diagne Gueye, Y. (2008) *Genre, Changements Climatiques et Sécurité Humaine: Les cas du Sénégal*, ENDA, Dakar

Edward, M., Satyanath, S. and Sergenti, E. (2004) 'Economic shocks and civil conflict: An instrumental variables approach', *Journal of Political Economy*, vol 112, no 4, pp725–753

Eldridge, C. (2002) 'Why was there no famine following the 1992 Southern African drought? The contributions and consequences of household responses', *IDS Bulletin – Institute of Development Studies*, vol 22, no 4, pp79–89

Enarson, E. (2000) *Gender and Natural Disasters*, IPCRR Working Paper no 1, International Labour Organization (ILO), Geneva

Hartmann, B. (2009) 'Don't beat climate war drums', *Climate Chronicle*, no 2, 9 December 2009, p8

IPCC (2007) *Climate Change 2007: Impacts, Adaptation and Vulnerability*, Contribution of Working Group II (WG2) to the Fourth Assessment Report of the IPCC, Summary for Policy Makers, IPCC, Geneva, pp7–22

Kok, M. T. J. and Jäger, J. (eds) (2009) *Vulnerability of People and the Environment: Challenges and Opportunities*, background report on Chapter 7 of the Fourth Global Environment Outlook (GEO-4), United Nations Environment Programme (UNEP) and Netherlands Environmental Assessment Agency, Nairobi, Bilthoven and The Hague

Lambrou, Y. and Piana, G. (2005) *Gender: The Missing Component in the Response to Climate Change*, FAO, Rome

Littlecott, C. (2007) 'Climate change: The global security impact', www.e3g.org/index.php/programmes/climate-articles/climate-change-the-global-security-impact/, accessed 10 July 2010

Malawi Government (2001) *Sexual and Reproductive Health Behaviours in Malawi: A literature review to support the situational analysis for the National Behaviour Change Interventions Strategy on HIV/Aids and Sexual and Reproductive Health*. National Aids Commission and Ministry of Health and Population, Lilongwe, Malawi

Mitchell, T., Tanner, T. and Lussier, K. (2007) *We Know What We Need: South Asian women speak out on climate change adaptation*, ActionAid International, Johannesburg and London

Neumayer, E. and Plümper, T. (2007) *The Gendered Nature of Natural Disasters: The Impact of Catastrophic Events on the Gender Gap in Life Expectancy, 1981–2002*, London School of Economics, University of Essex and Max Planck Institute for Economics, London

O'Brien, K. (2007) 'Commentary to the Paper of Úrsula Oswald Spring, Climate Change: A Gender Perspective on Human and State Security Approaches to Global Security', paper presented at International Women Leaders Global Security Summit, 15–17 November 2007, New York

Oswald Spring, Ú. (2007) 'Climate Change: A Gender Perspective on Human and State Security Approaches to Global Security', paper presented at International Women Leaders Global Security Summit, 15–17 November 2007, New York

Oxfam (2005) 'Gender and the tsunami', briefing note, March 2005. Oxfam, Oxford

Pan American Health Organization (PAHO, regional office of the World Health Organization) (1998) *Gender and Natural Disasters*, Women, Health and Development Program, factsheet, PAHO, Washington

Patt, A., Dazé, A. and Suarez, P. (2007) 'Gender and climate change vulnerability: What's the problem, what's the solution?', paper presented at the International Women Leaders Global Security Summit, 15–17 November 2007, New York

Schrader, E. and Delaney, P. (2000) *Gender and Post-Disaster Reconstruction: The Case of Hurricane Mitch in Honduras and Nicaragua*, World Bank Report, Washington

Skutsch, M. (2004) 'CDM and LULUCF: What's in it for women?', Note for the Gender and Climate Network (GenderCC), July 2004

Stern, N. et al (2006) *Stern Review on the Economics of Climate Change*, Cambridge University Press, Cambridge

UN (2007) UN Security Council/SC/9000, 17 April 2007, www.un.org/News/briefings/docs/2007/070404_Parry.doc.htm, accessed 12 January 2010

UNDP (United Nations Development Programme) (2007a) *Human Development Report 2007–2008: Fighting Climate Change: Human Solidarity in a Divided World*, Palgrave Macmillan, New York

UNDP (2007b) *Poverty Eradication, MDGs and Climate Change*, UNDP, New York

UNFCCC (United Nations Framework Convention on Climate Change) (2005) *Global Warming: Women Matter*, UNFCCC Conference of Parties (COP) Women's Statement, December 2005

UNIFEM (United Nations Development Fund for Women) (2005) *UNIFEM Responds to the Tsunami Tragedy – One Year Later: A Report Card*, UNIFEM, New York

WEDO (Women's Environment & Development Organizaton) (2007) 'Changing the Climate: Why Women's Perspectives Matter', factsheet, WEDO, New York

Wisner, B., Piers, B., Cannon, T. and Davis I. (2004) At Risk: Natural Hazards, People's Vulnerability, and Disasters, 2nd edition, Routledge, London

Wisner, B., Fordham, M., Kelner, I., Rose Johnston, B., Simon, D., Lavell, A., Brauch, H. G., Oswald Spring, Ú., Wilches-Chaux, G., Moench, M. and Weiner, D. (2007) 'Climate change and human security', simultaneously uploaded to Radical Interpretations of Disaster (Radix), www.radixonline.org/cchs.html; Disaster Diplomacy, www.disasterdiplomacy.org; Peace Research and European Security Studies (AFES-PRESS), www.afes-press.de/html/topical.html, accessed 12 January 2010

CASE STUDY 3.1

CLIMATE CHANGE AND WOMEN'S VOICES FROM INDIA

Biju Negi, Reetu Sogani and Vijay Kumar Pandey

Introduction

Be it up in the mountains of Uttarakhand in the central Himalayas or down in the plains of eastern Uttar Pradesh – in the already threatened ecosystems, made worse by climate change, it is the woman who is consistently the most vulnerable and yet the most determined not to give in. She is vulnerable because, here, she not only bears an inordinately larger portion of domestic and economic work at the household level (60–70 per cent; and up to 80–90 per cent in agriculture) (Kaushik, 2004), but also because of her lack of access and control over resources, and decision making power at the community level. And she is fiercely determined because, as a woman, as the keeper of her family and house, her struggles and her determination are her only respite. Characteristically, her concern and responsibilities towards her family (that includes the household cattle and animals) are what bring her face-to-face with climate change in her everyday life, and in a way that is neither felt nor understood by the social and political leaders nor even by her sons, brothers, husband, father, father-in-law and, ironically, at times even by her mother-in-law.

For the woman, climate change is at every turn. Because she is the natural resource manager, climate change confronts her through land erosion, changing and unpredictable rainfall patterns, floods, drought, degraded forests, polluted or depleted water resources, lost seeds and so on. Because she is the food and fodder manager, all these consequences of climate change lead to weaker agriculture, a longer trudge to the forests for an increasingly poorer fodder gathering, and ultimately poorer food for her family. This is the challenge that she is faced with day in and day out, 365 days a year. And yet, not for one moment does she lose sight of the reality and her holistic perspective, looking at climate change not in isolation but as part of the larger picture of marginalization – of people and their agriculture and environment.

Deep-rooted climate change

In a private conversation, Baisakhi Devi from Pauri Garhwal (Uttarakhand) said: 'The climate started changing very fast when they gave us poison [meaning chemical fertilizers] and new seeds to put in our fields, saying these will give us more produce. That is when everything started to change. We could not deal with it because we had lost our seeds and our soil had become weak, and so we had become weak.'

Indeed, all problems are related to one another, with each bearing upon the other. So, call it Green Revolution, deforestation, loss of agrobiodiversity and an even bigger loss of local wisdom and knowledge – or climate change – the repercussions and consequences are here and they are real.

The mountains always used to have winter rains and snow – as early as November up until January. But in the last 4–5 years, as Devaki of Nai village (district Nainital, Uttarakhand) says: 'Now it is snowing only towards the end of February and even then it stays for barely one day.' Snow in December–January meant that it stood and remained on the slopes longer and melted slowly to seep underground. Snow in February, with temperatures already warming, means it will melt quickly and run-off – leading to land erosion on the one hand and decreased moisture retention in the fields, on the other. Besides crops, this has affected the fruiting of apples and pears. The village of Aadukhan (district Nainital, Uttarakhand), which was known for its big, juicy peaches, now finds its fruit size getting smaller and also getting infected, because timely snow also prevented pest attacks.

Fruits are now flowering off-season. The flowers too are shrivelled, pollination scanty and fruiting weak, affecting both quality and quantity. At higher altitudes, *buraans* (rhododendron) is now flowering early, and so too is the fruiting of *kaafal* (a locally popular forest tree fruit). *Kaafal* picking in late April–May, for the women, was a social event when they would be relatively free from their agricultural chores and go to the forests in groups, and also meet there with other women from other villages. *Kaafal* picking was really a fun day, away from daily chores, to socialize, and gather fruit in addition. Now its early fruiting means the women are still in the middle of their agriculture work and cannot find enough time to go collecting. Indeed, much more than the physical impact, climatic change is wreaking social and cultural dislocation. Many of the local festivals, such as Phuldeyi and Navratri, were timed so as not to conflict with harvesting and sowing, which are now changing. Steadily, thus, the popular, age-old folk songs too may need to be changed and will eventually be lost.

In the plains of eastern Uttar Pradesh, the story is the same. Post-monsoon and before the winter sowing of wheat, the farmers take an intermediate crop – of mustard. According to Isravati Devi (Village Saarhe Khurd, District Sant Kabir Nagar, Uttar Pradesh):

> *In late September–early October, by which time monsoon has ended, we would sow mustard. But now, because rain may randomly and sporadically still be there or the flood waters have not receded entirely by that time (because of natural outlets plugged by embankments and roads), the mustard sowing is delayed. In soggy soil, its flowering too is poor. At the time of flowering, the weather needs to be slightly cold, but now it is still warm, and so there is less flowering and there are also pest and disease attacks.*

The same stands true for potato, which is a major food source for the people in winters. Adds Isravati: 'We would depend on potato for our winter food, but due to unseasonal warming or rain, the potato is also affected.'

Delayed mustard sowing affects the subsequent sowing of wheat, the main winter crop, as well. Wheat needs a sustained cold period to grow long enough before flowering. But, as Meera Devi of the village of Janakpur (Gorakhpur district, Uttar Pradesh) says: 'To ripen, wheat needs a consistently long period of cold. The winter dew also helps a lot by providing moisture to the crop. But in the last 2–3 years, this cold period has decreased and the warm temperature has affected our wheat crop. Also there is not much dew falling anymore and so we need to give an extra irrigation.' Meera rears some animals as well, but last year a very high summer temperature killed her three hens and three goats. She says: 'I have a buffalo also, but because of strong summer heat, its milk production has gone down by more than half – from 8 litres to 3–4 litres. The heat has also affected the fodder: one more reason for low milk yield.' Meera used to sell the milk but also retain some for use at home. 'But now, due to less milk, I lose as much as Rs60–70 [US$1.11–1.29] per day,' and her family too does not get to consume milk any more. The cycle of deprivation grows.

In the mountains, March–April is time for drizzles and hails. 'Small-sized hailstorms are good for soil moisture and also kill pests,' says Kaladevi of village Kuhn (Nainital, Uttarakhand), 'but now we get cricket ball size hailstones, which do the opposite and ruin everything.' Around the same time, in the plains of Uttar Pradesh, when it is time for mango flowering, for some years now, 'strong easterly winds are blowing and all the flowering falls. Because of this, the fruit size now is getting smaller and quantity is also less,' says Isravati, and adds: 'This definitely has affected our food and nutrition because for a month or two of summers, mango used to be a major source of our food and diet.'

For the woman in the rural India, this long chain of climate change impacts – from food insecurity to water scarcity – goes on to effect her health and nutrition. About 45.6 per cent of the women in Uttarakhand are anaemic (NCP, 2004) – because they are always the last to eat and also eat the least. The climate change impacts also end up adding to the pressure on women to provide and thereby increase their workload many times over. This is also taking into account that in Uttarakhand and eastern Uttar Pradesh, there is considerable male emigration (rapidly increased during the post-flood periods, in eastern Uttar Pradesh) due to marginalization of small farm agriculture. This adds to the number of woman-headed households, a term which only spells more work and responsibility, without any parallel and positive shift in recognition, respect or authority.

Women's responses

In the consequences of her increased workload and mental pressure, in the social trivialization of her labour, experience and knowledge, in the lack of accounting her work as a contribution to the local economy, a woman may sometimes (or even often) fail and her despair forces her to take anxious, destructive measures. Yet, not for one moment does she stop thinking and struggling.

In places where conventional, chemical intensive agriculture is practiced, with decisions made by men, the people have gone in for heavier doses of chemical fertilizers and pesticides – making the solution worse than the problem. This is even more so, where – with males migrating and sons looking for other employment avenues – family agriculture is being entirely given up and cattle sold off, leaving the woman to seek work as a labourer, with the task of spraying.

Or, because of early maturity and poor harvests, people are going in for extracting three crops per year instead of the normal two, which is not good for the fertility of the soil.

Or food grains have been replaced by vegetables and spices/condiments cultivation, which also bring in ready cash. While income is welcome, many women feel, replacing the entire food grain production is not a very wise decision in the long run. As an elderly Kamla Devi (village Aadukhan) put it: 'Ultimately, we have to go to the market – for rice and wheat. So, what is the point?'

But in areas where women still practise some modicum of traditional agriculture or where her local knowledge system is still alive and respected, or where organizations (such as Samudayik Chetna Kendra – Community Awareness Centre in Nainital district, Uttarakhand and Gorakhpur Environmental Action Group in eastern Uttar Pradesh) have sought to bring the people together on the issues of food and agriculture, women are responding to climate change with sustainable agriculture strategies and practices.

In the flood prone areas of eastern Uttar Pradesh, the women have worked to raise vegetable nurseries on raised ground. They have also gone in for cultivating other crops. Says Isravati Devi: 'We, small landholder farmers are no longer depending on single crop farming. In a situation of drought, we also cultivate maize and groundnut; but if there is a flood situation, we erect a platform on which we spread out vegetable vines. So at least harvest is not lost entirely, come what may.' In her area, community grain and seed banks have also been set up to distribute these to the needy in times of distress or as loans.

In Nai village (Nainital, Uttarakhand), when apples failed, people started planting lemon and other citrus fruit trees. In the mid-hills they have gone for a diverse range of tropical fruits, such as banana, guava and even mangoes (remember, it is warming!). In Bheerapani (Nainital, Uttarakhand) traditional crops, which are much more resilient to climate change effects, have been sought to be revived through seed and women's knowledge exchanges. Hemlata of village Khujethi (Nainital, Uttarakhand) said: 'This season we had a good yield of *mandua* (*Eleusine coracana*), while the other crops failed.' Village elder Hemanti of Sunderkhali (Nainital, Uttarakhand) added:

'We used to eat *mandua,* which is why we are still strong; not like the new generation which eats from the market and hence is weak!'

For her animals, Meera Devi now prefers to keep local, traditional species because 'like traditional crops, local animals are more resilient to changing climate situations.' Also, last year, in part of her one hectare field, she planted some mango trees and banana in a mixed formation. 'I do not know how climate change will affect my trees and banana. But we will see,' she adds with a ring of confidence in her voice.

Conclusion

For the women, addressing climate change means seeing it in the totality of environmental degradation and social disintegration. Addressing climate change also means recognizing their role, knowledge and strength as climate friendly, ecological farmers and not marginalizing their small scale agriculture, through policies promoting corporate agriculture.

Ultimately, it means not putting up any barriers to their collective strength and resilience, and demolishing the ones that inhibit their intrinsic capacity to innovate and solve.

Isravati Devi put it simply: 'I pray that when the government makes any plans for us, we must be consulted. Only then will the plans be effective.'

References

Kaushik, S. (2004) 'Situational analysis of women and girls in Uttaranchal', National Commission for Women, New Delhi, http://ncw.nic.in/pdf reports/Gender%20Profile-uttaranchal.pdf, accessed 18 January 2010

NCP (National Commission on Population) (2004) 'Presentation by Government of Uttaranchal on demographic profile', Government of India, http://populationcommission.nic.in/cont-en-ut.htm, accessed 18 January 2010

The authors wish to thank Leela (Joshuda), Hema (Haldwani) and Bhavan Singh (Chak Dalar), all Uttarakhand, for their help with this case study.

4

Cities, Climate Change and Gender: A Brief Overview

Prabha Khosla and Ansa Masaud

Introduction

Today more than half the world's population lives in urban centres (UNFPA, 2007). In the 1950s this was still 30 per cent and in 1975 just over 37 per cent (UN-HABITAT, 2009). Now, 52 per cent of the world's urban population resides in cities and towns of less than 500,000 (UN-HABITAT, 2009:11). Urbanization includes the growth of upmarket suburban areas alongside an increase of overcrowded tenement zones, ethnic enclaves, informal settlements and slums. Urban sprawl, socio-spatial challenges, lack of sustainable energy sources, insufficient serviced land, inadequate basic services infrastructure and unplanned peri-urban development are some of the consequences of this rapid growth. For countries of the south, how to manage urbanization continues to be an ongoing challenge.

One consequence of so much unplanned and rapid growth is the increasing vulnerability of millions of poor urban women, men and children to a changing climate, including rising sea levels, flooding, high winds and other climate induced hazards. The uneven distribution of vulnerability between and within urban areas and the way this is shaped by individual and household characteristics including age, gender, income-level, livelihood strategies, health-status, physical, economic and social assets demands a textured understanding of power, gender and intersectionality in urban areas, rather than a gender and diversity-blind lens. An intersectional analysis is critical to developing relevant, realisable and sustainable strategies for dealing with climate change in cities.

Cities and towns are home to millions of men and women, young and old in all their diversity and disparity, yet rarely does the discourse on climate change mention the need for action in urban areas. Neither do the words cities, climate change and gender often come together on the same page. For too many years, the dialogue has been about natural environments and rural areas. The discourse has been dominated by a

Table 4.1 *Global trends in urbanization (1950–2050)*

Region	Urban population (millions)					Percentage urban				
	1950	1975	2007	2025	2050	1950	1975	2007	2025	2050
World	737	1518	3294	4584	6398	29.1	37.3	49.4	57.2	69.6
More developed region	427	702	916	995	1071	52.5	67.0	74.4	79.0	86.0
Less developed region	310	817	2382	3590	5327	18.0	27.0	43.8	53.2	67.0
Africa	32	107	373	658	1233	14.5	25.7	38.7	47.2	61.8
Asia	237	574	1645	2440	3486	16.8	24.0	40.8	51.1	66.2
Europe	281	444	528	545	557	51.2	65.7	72.2	76.2	83.8
Latin America and the Caribbean	69	198	448	575	683	41.4	61.1	78.3	83.5	88.7
North America	110	180	275	365	402	63.9	73.8	81.3	85.7	90.2
Oceania	8	13	24	27	31	62.0	71.5	70.5	71.9	76.4

Source: UN-HABITAT (2009, p23)

dialogue that treats climate change as a technical matter disconnected from the realities of millions of people and the domain of power and social relations between genders. It needs to look at economic and income disparity, violation of human rights and disfranchisement, inequality and inequity, and institutional and political discrimination. The voices of millions of poor urban residents and movements need to be an integral part of this discourse.

Cities need to be seen as part of the solution for climate change. This chapter argues that the nexus of cities, gender and intersectionality and climate change needs to be on the front page of climate change deliberations and actions. It calls for a textured and multidimensional analysis of urban centres with a deliberate focus on poor women, men, children and their organizations, by national and local governments, civil society organizations, private sector institutions and international agencies.

Why cities are important to climate change debates and actions

Cities are the economic powerhouses of nations. They are centres of innovation, politics, finance, commerce, intellectual and cultural creativity and home to millions of urban residents. Cities are also centres of production, consumption and at the same time they can enhance innovation and advance clean energy systems, sustainable transportation, waste management and land use policies to reduce greenhouse gases (GHGs). On the other hand, most of the future population of countries will reside in cities and towns where the impacts of climate change are more likely to take place. Thus, cities and towns are inevitably key sites for actions to adapt and mitigate the harmful impacts of changing climatic regimes. Urban centres are particularly crucial for climate change actions due to their high populations and risk patterns which intensify climate change impacts. A discussion of the ways in which climate change might exacerbate humanitarian crises in urban areas must be predicated on the complex conjuncture of a number of underlying factors:[1]

- Urban areas are characterized by high population densities, which result in high exposures to disasters and climate change impacts.
- The combination of high vulnerability and exposure causes a high degree of risk.
- The majority of the world's urban population lives in coastal cities (including megacities such as Dhaka, Lagos, Shanghai and Mumbai); these cities are most at long-term risk of rising sea levels and storm surges associated with climate change and global warming.
- Typhoons, hurricanes, floods, tropical storms, landslides and forest fires are increasingly taking place in urban areas, causing severe damage; this has been witnessed, for example, in the latest stream of disasters in Indonesia, the Philippines, Haiti, Burkina Faso and Pakistan.
- In low and middle-income countries, the urban growth trend of secondary and smaller cities is proportionately faster than primate and megacities. Moreover, these same cities tend to have more limited investment in infrastructure – for example, water and drainage – weaker urban management and planning capability, and they attract less investment and donor interest. These factors make them predisposed to greater vulnerability to, but least prepared for, the environmental effects of climate change.
- Millions of poor urban women live in extreme vulnerability due to insecure tenure in informal settlements.
- Lack of appropriate land tenure and land use and building regulation controls for poor residents, particularly for women.
- Vital infrastructure is often lacking in high-density cities of the south.
- Urban areas represent a range of interlocutors, governance institutions and regulatory frameworks.

Cities host large concentrations of poor urban women and men. In cities of the south, poor urban women, men, young and old, constitute from 30–60 per cent of the urban population. Poverty increases risks and vulnerability to climate change impacts.

Furthermore, it is not only that cities will experience and feel the impact of intense weather events such as increased flooding, landslides, storm surges, but also that cities themselves and the processes of urbanization have created conditions that intensify such weather phenomena. Urbanization processes such as the removal of tree cover, paving of porous surfaces, the density of the built environment, the reclaiming of wetlands and river beds, the levelling of higher surfaces for construction, the loss of coastal vegetation, all contribute to increasing the intensity of flooding, landslides, storm surges and to creating urban heat islands.

Cities should also be a key focus for climate change discussions as climate impacts in rural areas are likely to trigger massive rural-to-urban migration as well as migration from the worst hit regions and countries to cities of neighbouring countries.

Gender, intersectionality and cities

In recent years, women's and gender rights advocates have called for a gender analysis in climate change negotiations as well as related processes, mechanisms and commitments. Organizing and lobbying efforts have all underlined the importance of gender equality and equity considerations in the United Nations Framework Convention on Climate Change (UNFCCC) processes if strategies for mitigation and adaptation are to be successful and sustainable.[2] Much of the debate has centred on natural environments, rural areas and livelihoods of rural women and does not reflect the realities of urban poor women and men. While the overall gender and climate change discussion is important, this dialogue excludes millions of poor urban women and men, young and old, from the discourse of climate change.

For improving the lives and the prospects of many poor urban women and girls living in slums in cities of the south, a gender analysis is critical to understanding how they live and work in cities. However, the gender analysis needs to be complemented with an understanding of other intersecting identities that can also marginalize them. Broadly speaking, urban populations tend to be more heterogeneous than rural populations and increasingly so in the cities of the south with their high rates of urbanization. The lives of millions of poor women and girls as well as men and boys are often an intersection of multiple identities and discriminations. For example, poor urban women and girls experience discrimination and violations of their human rights not only on the basis of their gender, but also because of other unequal power relations due to a combination of factors such as class, age, ethnicity, caste, race, aboriginality, sexuality, literacy, linguistic group, religious affiliation, status, ability/disability and geographical location.

The concept of intersectionality was introduced by Kimberlé Crenshaw in her seminal analysis of anti-discrimination legal precedents in the US which were informed by a 'single-axis framework... and contributed to the marginalisation of black women' (1989, p140). She unpacks court cases to demonstrate that anti-discrimination legislation was premised on feminist theory of white women's experience of discrimination and anti-racism premised on discrimination faced by black men. The lack of a multi-dimensional analytical framework undermined the efforts of working class black women in their struggles for equality and equity. According to Crenshaw contemporary feminist and anti-racist discourses have failed to consider the intersection of racism and patriarchy (1991, p1241), one could also add class, sexuality and location. Since then, numerous feminist scholars and organizations have added to this debate, demonstrating the need for a more nuanced analysis of the category 'women' (AWID, 2004; CRIAW, 2006). Needless to say, Crenshaw's research and analysis was particular to the US, but the reality of intersecting identities and multiple discriminations is also true for many poor women around the globe as witnessed by the organizing of, for example, black women in Brazil and Peru, by dalit women in India and indigenous women in countries of both the north and south.

A women's rights and gender intersectional analysis is essential to urban governance as it is to discussions, analysis, policies and actions on cities and climate change. Diverse women and girls are half the population of cities and poor women and girls are more than half the total urban poor population. And depending on the country and city under consideration, the poor are also likely to represent significant numbers of minority or ethnic women, men and children. Women and children (who are usually cared for by women), constitute the majority in many cities of the south. Furthermore, the majority of the elderly in many cities are women. Just plain demographics argue for a gender intersectional lens as a must for city planning and management. And even if there was no argument of demographics, women and girls have the right to equality and equity in cities as their basic human rights.

Additionally, much of the literature on cities and climate change and the poor does not unpack 'the poor' and rarely does the literature speak about poor women. The poor are often treated as an undifferentiated group (Satterwaithe, 2007; Douglas et al, 2008; Moser and Satterwaithe, 2008; ACCCRN, 2009). Clearly that is not the reality in most slums in the world. Socio-economic, ethnic, race, gender, religious and age stratification exists in all slums. The poorest of the poor are often women who are single parents, women-headed households from minority communities, or older single women and men, women and men with disabilities, or any combination of the above.

Using an undifferentiated lens to work with complex poor urban communities can exacerbate the marginalization and poverty of the most destitute.

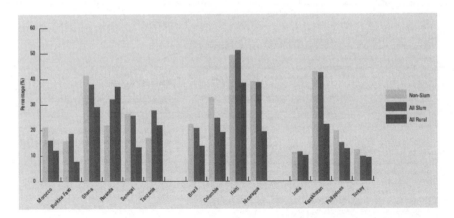

Figure 4.1 *Proportion of women-headed households in selected countries*

Source: UN-HABITAT (2006, p28)

Much of women's labour is in the care economy and for many poor women their income generation is often in the informal economy. Neither of these are waged nor valued as important contributions to family and society's well-being. This reality does not lead to the acknowledgement of poor women as important actors on the urban stage. Their living space is also their site of production and social reproduction. Their engagement in local community organizing for securing livelihoods, consolidating their living space, organizing services such as water, sanitation, health, education and food also makes them community activists and they are often the 'glue' of their communities, holding together social and survival networks. All these factors make diverse poor women key agents for inputs and innovation in securing their vulnerable neighbourhoods in the event of disasters. Since gender and social relations mediate how poor women and men live, work and navigate slums, women and men do not experience the physical and social spaces of their neighbourhoods in the same manner. This social and political stratification informs experience, knowledge and perceptions, and this is central to understanding and reducing the impacts of climate change events at the community, neighbourhood and slum level.

Understanding urban vulnerability

An understanding of urban vulnerability is critical to an understanding of why gender issues are an important consideration for climate change. Vulnerability is defined as an inability to cope with, resist or recover from shocks and stresses including climate change effects (Oxfam and

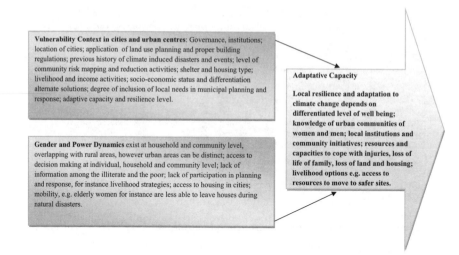

Figure 4.2 *Gender, urban vulnerability and adaptive capacity*

Source: Adapted from Oxfam and UN (2009, p18)

UN, 2009, p15). Poverty, gender, age and socio-economic differentials exacerbate vulnerability in urban areas. The following factors impact upon vulnerability of poor women and men in particular: (a) extreme overcrowding in slums and informal settlements with competition for few existing services; (b) unavailability of land or when available, it is in high risk areas and often lacks security of tenure; (c) the dependence of poor women and men on the informal market for livelihoods and income and insufficient financial services for the poor women in particular; and (d) extremely insufficient or a complete lack of infrastructure.

Power relations between the sexes, and gender roles in livelihood generation influence vulnerability and the adaptive capacity of individuals, households and communities.

Urban vulnerability is highly contextual as cities are in a constant state of flux. It is also impacted by population growth, city size, gender, marital status, class, ethnicity, income level, social safety nets, existing assets, site and type of housing and shelter, level of development in the slum/neighbourhood, security of tenure, level of infrastructure, financial resource investment, participation in planning and design of urban services.

Some settled areas may be vulnerable due to the nature of their location, for example, if they are in low-lying areas, or due to socio-economic reasons such as inadequate infrastructure, malnutrition and a lack of education for young girls and boys. Poorer urban households, which may be often headed by women, are usually at a higher risk due to weaker shelters, less safe locations and weaker resilience of infrastructure to withstand climate-induced damages. The vulnerability of the urban poor to floods is evident not only in the physical impacts, but also as they often experience increased rates of infectious diseases – including cholera, cryptosporidiosis and typhoid fever – and social stress and exclusion after flood events. During heatwaves, the very young, the elderly and people in poor health are particularly at risk due to their weak coping capacities. For example, in the event of floods and storms, households often face problems in getting timely and accurate information which can help them make decisions on relocation and preparedness.

Finally, it is important to emphasize that the vulnerability of urban dwellers to environmental hazards and humanitarian disasters is not just inherent to the physical conditions in which they live. Human disasters and vulnerability to disaster are socially constructed phenomena deriving from differential exposure to risk and preparedness, coping capacities and recovery capabilities.[3] Vulnerability in urban areas is differentially structured by the socio-spatial segregation of cities in which poor people are made more vulnerable by the lack of access to suitable land for housing, inadequate planning and environmental infrastructure, lack of effective disaster preparedness and mitigation policies and practices, poor governance and lack of empowerment (Zetter, 2009, p3).

How poor urban women, men and children currently manage the risks, assets, and resources in their neighbourhoods, i.e. their coping strategies, would be critical for adaptation strategies for cities. Local governments can be proactive in strengthening these capabilities as well as by providing much needed infrastructure for the provision of water, storm water drainage, sewerage, solid waste removal, safe energy and better housing.

Built environment, urban risks and climate change

Location is a major determinant of the type and frequency of natural hazards a city may experience (UN-HABITAT, 2009, p14). Eight of the ten most populous cities on the planet are located on earthquake faults, while 90 per cent of these cities are in regions vulnerable to destructive storms (UN-HABITAT, 2009, p14) as shown in Table 4.3. Coastal areas are densely populated and have large concentrations of economic activities.

Lack of urban and risk-sensitive land-use planning significantly increases vulnerability of poor women and men to climate-induced natural hazards. In many poor urban communities, homes, businesses and community facilities are situated on hazard-prone land, with unregulated,

Table 4.2 *Some climate change-related weather variables and extreme weather events and their impacts on cities and low-income residents*

Some climate change-related weather variables and extreme weather events	Impacts on vulnerable areas of the city	Impacts on poor urban women and men, girls and boys
Rise in temperature Heatwaves Drought Sandstorms	Severely reduced or complete loss of access to potable water. Heat Island effects. Unhygienic conditions. Desertification. Dust.	Increased susceptibility to dehydration for everyone but especially for babies, the young and older residents; collapse of hygiene conditions impacting babies, young and old and those with health considerations such as HIV/AIDS. Increasing risks from infections, sickness, disease and death for younger and older people due to inability to cope with the heat. Increase of respiratory diseases, such as asthma and bronchitis. More work for women, girls and boys to locate safe and affordable water; more work for women to look after sick relatives; girls and boys pulled out of school due to lack of water; menstruating girls drop out of school. Loss of income due to inability to do physically demanding work for men who are daily labourers, rickshaw pullers, gardeners, and so on. Loss of income for women from their home and water-based economic initiatives. Higher prices for water, fuel and food which in some places are borne by women; loss of water curtails some religious and cultural practices. In-migration to towns and cities or out-migration from dryland cities. Men are likely to migrate out and leave women and children behind. Reduction of the supply of food to urban areas and price increases causing malnutrition, particularly for girls and (older) women; lack of food can increase tensions and rifts between vulnerable groups trying to meet their basic needs.

Table 4.2 *(continued)*

Some climate change-related weather variables and extreme weather events	Impacts on vulnerable areas of the city	Impacts on poor urban women and men, girls and boys
High winds Heavy rains Flooding Saline intrusion Cyclones Hurricanes Typhoons Storm surges Sea-level rise Coastal erosion Landslides	Damage to and loss of shelter. Damage to and loss of urban infrastructure: electricity, water, roads, railways, sewage systems etc. Damage to and loss of public spaces, including community halls, religious structures and collectively managed infrastructure. Damage to and economic loss to homes, productive assets and belongings. Contamination of ground water, wells, and other water sources. Overflow of drainage systems. Severe damage to coastal communities.	Women and their children living in shacks or informal settlements near rivers/coastal areas are highly vulnerable as unregulated and poor quality shelters often collapse causing injuries and deaths. Injuries and loss of life, esp. of young children or older family members; trauma and stress for women to protect children, household members, goods, and secure water and fuel. Increased risk of water-borne diseases such as diarrhoea, malaria, cholera, yellow fever, especially for babies, young children and the sick and old, increasing women's workload. Loss of income for women and men. Women and older women and men who have no or irregular income, or are sometimes supported by community structures are further impoverished. Further impoverishment due to loss of shelter, productive assets and belongings of women and men. Loss of workshop/tools for both women and men; destruction of women's vegetable patches/gardens; loss of food and insecurity with more pressure on women to secure food. Increase in the cost of water and food which will be especially difficult for women-headed, and single and older women households. Inter-city and intra-city displacement, loss of land and property for those that have security of tenure. Migration to other parts of the city or other towns and cities.

Source: The development of this table was informed by Wilbanks et al (2007) and UNFPA (2009, pp39–51)

Table 4.3 Ten most populous cities and associated disaster risk (2009)

Cities	Population (in millions)	Earthquake	Volcano	Storms	Tornado	Flood	Storm surge
Tokyo	35.2	✓		✓	✓	✓	✓
Mexico City	19.4	✓	✓	✓			
New York	18.9	✓		✓			✓
São Paulo	18.3			✓		✓	
Mumbai	18.2	✓		✓		✓	
Delhi	15.0			✓		✓	
Shanghai	14.5	✓		✓		✓	✓
Kolkata	14.3	✓		✓	✓	✓	✓
Jakarta	13.2	✓				✓	
Buenos Aires	12.6			✓		✓	✓

Source: UN-HABITAT (2009, p14)

unsafe construction and inadequate services. It is estimated that more than half of the population in urban areas is living within 60km of the sea, while 3–4 of every 10 non-permanent houses in cities in developing countries are located in areas prone to floods, hurricanes, tropical cyclones, landslides and other natural disasters, further exacerbated by climate change (UN-HABITAT, 2008a). Climate change brings new challenges, which impact on the natural and built environments and aggravate existing environmental, social and economic problems. These events affect different aspects of spatial planning and the built environment, including the buildings, infrastructure, community cohesion and physical assets of poor women and men. Poor communities living in these conditions cannot afford savings or asset accumulation, and their vulnerability is immense. Furthermore, more than 1 billion people across the world are already living in urban slums where many are exposed to urban hazards and associated vulnerabilities. Alongside these, other groups such as refugees, internally displaced persons, returnees and stateless persons, also share densely populated and poorly serviced communities, making them more vulnerable to the impact of disasters. This increases the risk that disasters will devastate both the built environment and the social economy, resulting in longer-term and more extensive setbacks to recovery and development.

The majority of poor urban women, men and children live in the most vulnerable parts of urban areas, such as in informal or 'illegal' as well as regularized slums in cities located in, near or on wetlands, landfill sites, river channels, river edges, flood plains, rocky outcrops, steep slopes and directly on coasts. What compounds their vulnerability in these areas is that their homes are made of impermanent materials and so cannot withstand shocks, as well as their vulnerability due to the lack of infrastructure such as water, sanitation, sewerage and drainage and insufficient services such as schools, hospitals, transportation and electricity.

Gender responsive land-use planning, preparedness plans and inclusive city planning and governance are highly significant approaches to mainstream disaster risk reduction and climate change related risks into urban development processes and reduce the risk of natural disasters due to changes in climate. Land-use planning provides opportunities to partner with local actors for risk mapping and community resilience building. This includes partnerships between national and municipal governments, planners, development agencies, bilateral partners, women and community groups, community-based organizations (CBOs), non-governmental organizations (NGOs) and the private sector. In order to cope with the effects of climate change through rising sea levels, cities will need to implement innovative gender-responsive and women-centred adaptation and mitigation strategies by making disaster resilient infra-structure, disaster resistant shelter and housing, clearing disaster prone areas through comprehensive city-wide planning with adequate services and infrastructure in both pre- and post-disaster settings. City management and governance needs to be approached within a holistic framework which responds to the impacts of disasters, climate change, inequality, poverty and vulnerability. For instance, city examples show that making gender-responsive and storm and hurricane resistant housing and shelters in the reconstruction phase is an opportunity to reduce the impact of natural hazards in future. However, the lack of representation of poor women in local and sub-regional formal decision-making structures (for example, at provincial, community and city level and programme implementation) means that poor women's interests are not being adequately considered in risk reduction, disaster and climate change response and post disaster planning.

Improving disaster response and reducing risks

Enhancing preparedness

The strengthening and empowering of communities and organizations in urban and peri-urban areas is instrumental for poor women and men to map their risks, identify hazards, develop preparedness plans and enhance and build capacity to mitigate the impacts of climate-induced disasters. For communities to sustain the impact of climate change, the following minimum actions need to be undertaken: first, reconsider urban planning from a poverty and gender-responsive lens. Second, build the capacities of local communities to respond to the needs arising from climate-induced disasters. Third, support local communities to identify the hazards and risks they face, as well as their coping strategies, then translate this knowledge to local community action plans for mitigation or reduction to these hazards. Lastly, create effective financing partnerships between the national government, donor countries, international agencies and communities to address the impact of gender and climate change at the local level. There needs to be an incremental approach to addressing climate change at the city level, and this needs to be embedded in national and city sustainable development frameworks.

Box 4.1 *Nepal: Women reducing risk through community upgrading*

The Pragati Women's Cooperative, one of several grassroots women's cooperatives partnering with Lumanti Support Group for Shelter, is a Kathmandu-based NGO, established in 1993, to reduce risk from climate change events in Nepal. The Pragati Mahila Uthan Saving & Credit Cooperative represents squatter women of the Kathmandu Municipality in Nepal and is working to reduce women's dependency on moneylenders. The cooperative provides women with access to and control over financial resources and more than 350 members have taken loans for various income generating activities. The cooperative is owned and managed by 440 women from 29 saving groups belonging to 10 communities. The cooperative has a total savings of Rs 9 million (almost US$123,000) and has outstanding loans worth Rs10.52 million (US$143,742). In addition to the compulsory monthly savings this cooperative boasts ten savings products ranging from housing savings, children's savings to festival savings and fixed deposits. Various loan products with different interest rates and repayment periods consist of products such as high-interest loans, purchase of land for building a house, and so on. For emergencies, most members turn to their local savings and credit groups who ensure that they keep at least Rs1500–2000 (US$20.50–27.33) available at all times in case of emergencies. The cooperative provides large and small loans to different communities to improve community infrastructure and basic services affected from natural disasters. For example, Rs8000 (UD$110) was given to the Chandol community to cover open drains, Rs5000 (US$68) to Hattigauda to pay for a water connection and Rs2000 (US$27.33) each to Kadya Bhadrakali to repair their community buildings and to Tikuri to construct toilets. The cooperative essentially handles all the financial needs of its members.

As the women from Pragati demonstrate, when women are organized in groups, federations or cooperatives they have a greater chance of impacting decision making processes, influencing disaster risk reduction programs and accessing resources.

Urban risk reduction and resilience building measures that are embedded in the community development efforts of the Network of Women's Cooperatives, Mahila Ekta Samaj (Federation of Women's Groups) and the Baso Bas Basti Samrochan Samaj (Federation of Squatter Communities), both supported by Lumanti, are proving to be unique models in community infrastructure upgrading. The squatters federation continues to focus on land tenure issues, and the women's federation on savings and credit for income generation – but is increasingly tackling issues of housing, infrastructure and land tenure. The settlement committee in Narayantol, a small squatter community in a hilly part of Kathmandu, identified the need to prevent landslides in their settlement and worked together to create paved stairways and build retaining walls along the landslide prone settlement. The squatter community identified sanitation as a priority issue in need of improvement, as the lack of toilets and adequate drainage and sewage in the settlement was making the community – and particularly women – vulnerable to both climate-related floods as well as health problems. The Narayantol community has also recently installed a low cost up-flow bio-filter unit for sewage treatment which has been drawing a stream of visitors who want to learn from this community. In Kharipakha, a 150-household squatter settlement in Kathmandu, communities decided to build a paved cover over the stream as it was being used as a sewage disposal area. This intervention has helped to reduce the water-borne diseases in the settlement and enabled young people to safely engage in outdoor activities. In Bharatpur, fast growing municipalities in Nepal, women from the squatter communities have been taking the lead in the construction of 750 toilets. Starting with a basic hygiene assessment of communities, they go on to mobilize and motivate families to build toilets and then supervise the construction of these toilets.

Source: Adapted from Disaster Watch (2009)

Building resilience

A disaster and climate-resilient city is one that exhibits capacities to adapt when exposed to natural hazards in order to minimize damage and losses. Building urban resilience will require improving urban risk and capacity mapping, mapping of natural resource bases, urban infrastructure, creating more effective, gender-sensitive and pro-poor structures of governance, and increasing the capacity of individuals and communities to address these new challenges. These actions will also identify local hotspots and assist in setting up of early warning systems at the municipal level. For building resilience, the dialogue on climate change impacts needs to engage urban planners, architects, land-use planners, shelter experts, ministries for planning, housing and land, food security and emergency response departments, and the communities themselves. Building resilience relies strongly on the availability of resources and assets, quality of local knowledge and of local capacity and willingness to act. Urban resilience can also be facilitated through the adoption of pro-poor, gender-sensitive strategies that enable individuals and households to develop sustainable and resilient livelihoods such as diversifying sources of income for extremely poor groups in different seasons of the year. Indeed, having a strong and diverse economic base during the course of the year is one of the main factors that can help households to cope with the shocks and stresses that are likely to become more frequent in the long term.

Conclusion: Taking on the urban climate challenge

In the years since the Rio Earth Summit (1992), the notion of sustainable cities has shifted the urban planning paradigm. The linkages between the natural, socio-economic and political environments of cities are much better understood in terms of designing and managing the urban fabric than in previous years. The potential and existing scale of the destructive impacts of climate change on cities almost requires an acceleration of the holistic planning implied in sustainability planning.

In spite of the above, even the shift to decentralization, 'good governance' and sustainability in cities has not sufficiently been informed by pro-poor and gender-sensitive criteria. Research, studies and policy recommendations that address women, gender and poverty in cities now present a sizable body of literature. This material covers a wide-spectrum of issues such as urban planning, politics, participation and decision-making, governance, gender mainstreaming of municipal corporations, policies and programmes, decentralization, local economic development, transportation, housing, water and sanitation, recreation, paid and unpaid work, violence against women and girls, safer cities, and post-disaster reconstruction (RTPI, 2001; CEMR, 2006; Fainstein et al, 2005; Morna and Tolmay, 2007; UN-HABITAT, 2008b, c[4]). This wealth and experience needs to inform urban governance and adaptation planning for climate change in cities.

Adaptation strategies offer one means of facilitating a textured analysis into city management. Measures such as hazards mapping and assessment, risk reduction and management, vulnerability and capability assessments, and early warning systems are all critical elements of adaptation planning. These measures will only be effective if they are informed by and developed in the context of the most vulnerable residents of cities and especially with the participation of the many poor women, men, boys and girls. Furthermore, adaptation strategies need to be aligned with the existing urban governance and sustainability planning processes as well as national development planning frameworks.

Recognizing the scale of adaptation efforts needed in many cities and towns particularly of the south, pro-poor and gender-sensitive adaptation planning offers an opportunity to bring together the development agenda of equality, equity and justice with climate change imperatives.

Some key points to further this inclusiveness include:

- Shift the focus of the global discussion on climate change away from a primarily technocratic exercise to one employing the language of global justice and human rights, including the right to development and gender equality and equity. This is more than a semantic shift; it acknowledges that a north-south deal on climate change can only be completed when it incorporates a financial agreement that considers the questions of equity and fairness on a par with the need to reduce GHG emissions fast. In such a pro-poor climate-just deal, gender equity will have to feature prominently (adapted from Schalatek, 2009, p24).
- Integrate pro-poor and gender-sensitive criteria into planning, design, implementation, monitoring and evaluation of hazards mapping and assessment, risk reduction and management, vulnerability and capability assessments, early warning systems and contingency planning.
- Data collection should be disaggregated by sex, ethnicity, age and other social relations and should inform adaptation planning.
- Adaptation strategies should focus on the most vulnerable areas of the city. Resources allocated to these strategies need to ensure the equal participation of poor women, men, boys and girls to bring their knowledge of risk management into adaptation strategies. Local knowledge is key to creating sustainable neighbourhoods.
- Allocate adequate resources to address the needs of women in climate change mitigation, adaptation and disaster risk reduction, for example through funding appropriate and environmentally sound technologies and supporting poor women's initiatives in managing slums and informal settlements (adapted from Manila Declaration, 2008).
- Resource streams from the climate change mechanisms need to be channelled to build the capacities of both local governments in terms of managing cities under risk as well as the vulnerable communities

themselves in enhancing their capabilities and coping strategies. Capacity-building resources need to facilitate the engagement of poor women as equal decision-makers and knowledge partners.

• Develop a set of gender-sensitive criteria for all new climate finance mechanisms supporting adaptation, mitigation, capacity-building and technology transfer. This includes the funds administered under the UNFCCC and the Global Environment Facility (GEF) as well as the Climate Investment Funds (CIFs) and bilateral funds. Additionally, these funding streams should be accessible to local governments and organizations of the urban poor including women's groups in low-income neighbourhoods (adapted from Schalatek, 2009, p24).

Focusing on the points above will bring the challenge posed by the agenda of equality, equity and justice in urbanization and climate change within the realm of the possible.

Notes

1   Part of the factors mentioned are adapted from the Inter-Agency Standing Committee (IASC) Paper on Climate Change, Urbanization and Humanitarian Crises, developed for the IASC Task Force on Meeting Humanitarian Challenges in Urban Areas (2009).
2   See the statements and press releases from the Global Gender and Climate Alliance (GGCA), GenderCC and the Women's Environment & Development Organization (WEDO) (2009, 2010); for more information, see Chapter 8.
3   Hurricane Katrina in New Orleans, and its aftermath, provided dramatic evidence of the socially constructed nature of disasters. See also Chapter 5, Case study 5.7.
4   See also the work of the Gender and Disaster Network (GDN, www.gdnonline. org), Global Land Tools Network (GLTN, www.gltn.net/en/e-library/gender/ index.php) and the Centre on Housing Rights and Evictions (COHRE, www. cohre.org/women).

References

ACCCRN (Asian Cities Climate Change Resilience Network) (2009) 'Cities and resilience', Climate Policy Brief, www.rockefellerfoundation.org/news/publications /new-reports-climate-change-resilience, accessed 10 July 2010
AWID (Association of Women in Development) (2004) 'Intersectionality: A tool for gender and economic justice', *Women's Rights and Economic Change*, no 9, August, Toronto, www.awid.org/eng/Issues-and-Analysis/Library/Intersectionality-A-Tool-for-Gender-and-Economic-Justice/%28language%29/eng-GB, accessed 20 January 2010

CEMR (Council of European Municipalities and Regions) (2006) *European Charter for Equality of Women and Men in Local Life*, www.ccre.org/news_detail_en.htm?ID=879, accessed 20 January 2010

Crenshaw, K. (1989) 'Demarginalizing the intersection of race and sex: A black feminist critique of antidiscrimination doctrine, feminist theory and antiracist politics', *University of Chicago Legal Forum*, vol 1989, pp139–167

Crenshaw, K. (1991) 'Mapping the margins: Intersectionality, identity politics, and violence against women of color', *Stanford Law Review*, vol 43, pp1241–1279

CRIAW (Canadian Research Institute for the Advancement of Women) (2006) *Intersectional Feminist Frameworks: An Emerging Vision*, Ottawa, ON, Canada, www.siyanda.org/search/summary.cfm?nn=2930&ST=SS&Keywords=gender&SUBJECT=0&Donor=&StartRow=321&Ref=Sim, accessed 20 January 2010

DisasterWatch (2009) 'DRR initiatives in Nepal', DisasterWatch, www.disasterwatch.net/resources/DRR-Nepal.pdf, accessed on 24 February 2010

Douglas, I., Alam, K., Maghenda, M., Mcdonnell, Y., Mclean, L. and Campbell, J. (2008) 'Unjust waters: Climate change, flooding and the urban poor in Africa', *Environment and Urbanization*, vol 2008, no 20, p187

Fainstein, S. S. and Servon, L. J. (eds) (2005) *Gender and Planning: A Reader*, Rutgers University Press, New Brunswick, NJ

Manila Declaration for Global Action on Gender in Climate Change and Disaster Risk Reduction, October 2008, www.wedo.org/learn/library/media-type/pdf/manila-declaration-for-global-action-on-gender-in-climate-change-and-disaster-risk-reduction, accessed 4 January 2010

Morna, L. C., and Tolmay, S. (eds) (2007) *At the Coalface: Gender and Local Government in Southern Africa*, Gender Links, Johannesburg

Moser, C. and Satterwaithe, D. (2008) 'Towards pro-poor adaptation to climate change in the urban centres of low- and middle-income countries', Climate Change and Cities Discussion Paper no 3, first prepared for the World Bank's Social Development Department Workshop on the Social Dimensions of Climate Change, 5–6 March

Oxfam and UN (2009) 'Responding to climate change in Vietnam: Opportunities for improving gender equality', policy discussion paper, UN and Oxfam, Hanoi

RTPI (Royal Town Planning Institute) (2001) *Gender Equality and Plan Making: The Gender Mainstreaming Toolkit*, RTPI, London

Satterwaithe, D. (2007) *Climate Change and Urbanization: Effects and Implications for Urban Governance*, UN/POP/EGM-URB/2008/16, 27 December 2007

Schalatek, L. (2009) *Gender and Climate Finance: Double Mainstreaming for Sustainable Development*, Heinrich Böll Foundation North America, Washington, DC

UNFPA (United Nations Population Fund) (2007) *The State of the World Population 2007*, UNFPA, New York

UNFPA (2009) *The State of the World Population 2009: Facing a Changing World: Women, Population and Climate Change*, United Nations Population Fund (UNFPA), New York

UN-HABITAT (United Nations Human Settlements Programme) (2006) *State of the World's Cities 2006/7: The Millennium Development Goals and Urban Sustainability: 30 Years of Shaping the Habitat Agenda*, UN-HABITAT, Nairobi

UN-HABITAT (2008a) *State of the World's Cities 2008/9 : Harmonious Cities*, UN-HABITAT, Nairobi

UN-HABITAT (2008b) *Gender in Local Government: A Sourcebook for Trainers,* UN-HABITAT, Nairobi

UN-HABITAT (2008c) *Cities and Climate Change Adaptation,* UN-HABITAT Donors Meeting, Seville

UN-HABITAT (2009) *Global Report on Human Settlements 2009: Planning Sustainable Cities,* UN-HABITAT, Nairobi

Wilbanks, T. J., Romero Lankao, P., Bao, M., Berkhout, F., Cairncross, S., Ceron, J. P., Kapshe, M., Muir-Wood, R. and Zapata-Marti, R. (2007) 'Industry, settlement and society', in M. L. Parry, O. F. Canziani, J. P. Palutikof, P. J. van der Linden and C.E. Hanson (eds) *Climate Change 2007: Impacts, Adaptation and Vulnerability. Contribution of Working Group II to the Fourth Assessment Report of the Intergovernmental Panel on Climate Change,* Cambridge University Press, Cambridge, p357–390

Zetter, R. (2009) 'Climate change, urbanization and humanitarian crises', draft paper to the Inter Agency Standing Committee task force on Meeting Humanitarian Challenges in Urban Areas

Case Study 4.1

Mitigation of Greenhouse Gases (GHGs) by Informal Waste Recyclers in Delhi, India[1]

Prabha Khosla and Bharati Chaturvedi

India is the third largest aggregate emitter of greenhouse gases (GHGs) in the world, after China and the US (Government of India, 2009). According to the World Resources Institute (WRI), in 2005 India's emissions from the six major GHGs amounted to 1853 million tonnes of CO_2 equivalent (TCO_2e), which represents nearly 5 per cent of the global total. By way of comparison, China and the US account for roughly 19 per cent and 18 per cent of global emissions respectively. Because of India's large population and development path, per capita emissions remain relatively small; whereas developed countries such as the US, Canada and Australia all emit more than 20 TCO_2e per capita per year, Indians emit 1.7 TCO_2e per person per year (WRI, 2009).

In many countries of the developing world, the urban poor form the backbone of recycling programmes. Informal waste pickers, waste recyclers and small junk dealers, collectively known as the 'informal recycling sector', make up as much as 2 per cent of the urban population in Asia and Latin America (Wilson et al, 2009). These are men, women, and children who forage through rubbish heaps and depend on the revenues derived from selling recovered materials for all or part of their livelihood. Their work provides sanitation services to the municipalities where they live and results in reductions in GHGs.

According to the Municipal Corporation of Delhi, current daily waste generation is roughly 8500 metric tonnes (*Hindustan Times*, 2009). Independent analysis by Chintan Environmental Research and Action Group revises this figure upward, in the range of 9000–10,000 metric tonnes per day. Chintan projects that by 2020, municipal solid waste in Delhi will swell to 23,000 metric tonnes per day (Chintan, 2003). Chintan's new study, *Cooling Agents*, documents the role and impact of informal waste recyclers in mitigating GHGs. The study estimates that waste pickers collect 15–20 per cent of Delhi's total waste by weight and recycle virtually all possible recyclable materials they can. By recycling glass, metals, plastics and paper alone, the informal sector in Delhi reduces emissions by an estimated 962,133 TCO_2e each year. This is roughly equivalent to removing 176,215 passenger vehicles from the roads annually or providing electricity to about 133,444 homes for one year (US estimates).

Urban informal waste collectors and recyclers are the poorest of the poor. In Delhi, they are organized into a federation called Safai Sena – the Cleaners Army. The Safari Sena is mostly composed of lower caste women and men and Muslims of both sexes. In Delhi, the informal recyclers are 25 per cent women, 20 per cent children and the remaining 55 per cent are men. In most Indian cities women and men are both involved in the informal recycling sector in large numbers. This is not the case for Delhi. In Delhi, the lower number of women in informal recycling can be attributed to four major factors. First, due to their household and child raising responsibilities, women cannot roam far from their homes to collect waste. It is difficult for them to take young children and babies along with them. They are too poor to pay someone to look after their children while they work. Secondly, the increasing wealth of the upper middle classes of Delhi has created a greater demand for domestic workers and it is easier for women than for men to get jobs as domestic workers. Furthermore, domestic work is not as physically demanding as waste collection and considered more prestigious than scavenging for recyclables. Thirdly, a significant number of the Hindu men have left their wives in rural areas to protect their tiny landholdings or other assets. Fourthly, waste collection in Delhi requires lifting and carrying around large quantities of weight. Women complain that this it too much for them to carry, even in two rounds, but they accompany family members. Primarily poor Muslim and lower caste men carry the wastes and recyclables. Women are mostly involved in waste segregation, and do it as a part-time activity along with minding children and the demands of household responsibilities. For men to earn a minimum of Rs150 a day (US$3.31) requires obtaining and sorting more than 60kg of waste.

The informal recycling sector drives the city's recycling efforts, keeps the streets clean, and saves civic agencies huge sums of money. Additionally, informal recyclers are mitigating huge quantities of

GHGs. Despite such enormous benefits to cities and society, informal recyclers continue to face harassment from the police and civic authorities, live and work in abysmal conditions, are still the poorest of the poor, are considered illegal and a public nuisance, and do not have designated working spaces in the cities.

Cities could benefit enormously by working with federations of waste recyclers and reduce poverty and GHG emissions at the same time as enhancing urban sustainability and justice. Cities can be proactive by recognizing the informal waste recyclers as an integral part of the municipal waste management systems of urban centres. They can create and provide designated areas for the collection and recycling of recyclables which can be managed by the recyclers themselves, thereby increasing job security and the income generation potential of poor men and women. Recently, the New Delhi Municipal Corporation launched a pilot project in partnership with Chintan, that targeted to help formalize waste pickers' work. This new partnership also opened the door to a discussion with the Medical Officer of Health on the very critical issue of workers' health and safety.

Notes

1   This case study has borrowed liberally from 'Cooling Agents: An Analysis of Climate Change Mitigation by the Informal Recycling Sector in India', available at www.advocacynet.org/files/Cooling%20Agents%20full.pdf; *The story of Chintan* by Bharati Chaturvedi; and from an interview with Bharati Chaturvedi, the Director of Chintan.

References

Chintan (2003) *Space for Waste: Planning for the Informal Recycling Sector*, Chintan, New Delhi

Government of India (2009) *The Road to Copenhagen: India's Position on Climate Change Issues*, Government of India, New Delhi

Hindustan Times (2009) 'Time to "bale" out', 31 May 2009

Wilson, D. C., Araba, A. O., Chinwak, K. and Cheeseman, C. R. (2009) 'Building recycling rates through the informal sector', *Waste Management*, vol 29, no 2, p629

WRI (World Resources Institute) (2009) Climate Analysis Indicators Tool (CAIT) version 6.0

CASE STUDY 4.2

GENDER MAINSTREAMING IN THE CLIMATE CHANGE RESPONSE OF SORSOGON CITY, THE PHILIPPINES

Bernhard Barth

Sorsogon City, one of 120 cities of the Philippines, has a land area of 313km^2 and a population of 151,454 (as of 2007), growing at a rate of 1.78 per cent annually. In 2000, women in the city accounted for 49.7 per cent of the population. The 2007 census does not present sex-disaggregated data. Sorsogon's economy is based mainly on agriculture, fishing, trade and services. Sorsogon City, the capital of Sorsogon Province, is the administrative, commercial, and educational centre of the province.

In 2006, two super typhoons caused widespread devastation within a two month interval, leaving in their wake a total of 27,101 families affected and 10,070 destroyed houses. The first typhoon, in just five hours, caused damages to public infrastructure estimated at 208 million pesos or US$4.3 million. While the city was used to typhoons, their ferocity and rapid succession led to the popular perception that this was a sign of climate change. In August 2008, the mayor launched the city's climate change initiative, requesting the technical support of the United Nations Human Settlements Programme (UN-HABITAT).

Until then, the common idea was that, while climate change affected the city, this was a global and national issue requiring limited action from the local government. The mayor initiated a series of briefings for decision-makers and community leaders to enhance the basic understanding of climate change and the important role of local government. It was agreed that climate change vulnerability and adaptation assessments as well as a GHGs emissions inventory would be conducted by the city itself, primarily because funding for external consultants was not available. The assessment was to provide the city with the required information upon which a climate change plan could be developed.

Initially the city was particularly keen on getting reliable climate predictions and was hoping for a detailed localized climate model. However, it became clear that localized models would only be available by mid-2009 – a date that has been pushed back several times since then. For the region (Bicol, of which Sorsogon is part) as a whole, it was projected that more cases of prolonged monsoon rains causing rainfall exceeding 2800–3500mm per year were likely and that more extreme weather events were also likely to occur.

The assessment was based on the national guidelines where vulnerability (in accordance with the Intergovernmental Panel on Climate Change) is defined as a function of sensitivity, exposure and adaptive capacity. For a better understanding of the 'sensitivity' it was agreed that local weather and sea level observations were to supplement the available national and regional (subnational) climate change models. It was further agreed that the participatory methodology developed by the Sustainable Cities Programme was appropriate to triangulate the background information on climate sensitivity with community observations. The participatory methodology was the primary entry point to understand local exposure to weather-related hazards and to understand adaptive capacity. Local radio and TV stations, schools, youth and community groups were happy to be associated with the programme and actively participated in raising awareness with regard to climate change impacts and possible responses.

During consultations, residents recounted how typhoons and storm surges over the last decade had become stronger and more destructive. These personal accounts were recorded as evidences of climate change impacts. Using hand-drawn maps, local women and men graphically described the changes in the reach of tidal flooding and identified the areas gradually lost due to sea level rise and erosion. This participatory exercise promoted ownership by the women and men of the community in the assessment process and results and increased their awareness of climate change impacts. Moreover, the process empowered the people to work together with the local government in finding practical solutions that they can personally act on.

At this stage, a clear commitment to a participatory approach was prevalent. However, the understanding of and commitment to gender analysis and gender mainstreaming was less strong and to some extent imposed by UN-HABITAT. A draft of the vulnerability assessment states:

> *[Focus Group discussions] with communities revealed that in previous disasters… women experienced heavy burdens in that they needed to extend their roles to cope and recover from the damages in their homes and livelihoods. Women in Sorsogon City during the past two cyclones and in the context of disaster recovery expanded their roles to generate additional income to support the family. Immediately after the cyclone, the women were in the forefront of looking for resources that could be used to restore or augment their limited and damaged livelihoods. They tried accessing financial resources support and small business information and training programs from local micro-finance organizations in the city… Indeed, the women in Sorsogon played an important role in the overall livelihood and social recovery after the two super typhoons that devastated the city in late 2006. (UN-HABITAT, 2009)*

The draft report concludes that female-headed households in particular were vulnerable to the impacts of climate change and that a climate change response would provide the city with a 'great opportunity [for] mainstreaming equitable and responsive gender programmes'.

A validation workshop, bringing together city and community leaders as well as representatives from faith-based and other civil society organizations, discussed the findings of the vulnerability assessment. Women were well represented in the workshop and participated actively. Climate hotspots (flooding, landslides and storm surge impacts) were identified and eight Barangay (lowest level of local government) with a high number of low-income households were identified as potential sites for the climate change programme. The validation workshop concluded with the request to conduct more research on the social dimension of climate change vulnerability, to develop robust criteria for the prioritization of Barangay and to develop propositions for action which were to be discussed in the city consultation. The revision of the vulnerability assessment addressed gender concerns in more detail and the criteria for priority setting explicitly contained gender indicators.

All households in those Barangay which were regarded as containing the most vulnerable communities participated in the second level assessment. For the first time the city and the community had detailed information on poverty and social indicators which were visualized in simple maps.

Key survey questions related to the impacts of climate change on the households – with a clear identification of the households as female or male-headed. The initial assessment had clearly indicated that the most vulnerable households were female-headed. Furthermore, it had become clear that women's burdens were exacerbated in a number of areas after the disasters and that the survey questions needed to include indicators to monitor these.

The indicators include: household expenditure, breakdown of monthly expenditure, child labour, food provision for household members, type of sanitation facilities, access to improved water sources, underweight children, persons with disabilities, gender disparity in primary and secondary education, access to secure tenure and households that can afford repairs to their houses after natural disasters. The survey focuses on households, yet sex-disaggregation is consistently undertaken where possible.

The city government, local communities and the private sector are now engaged in planning for climate change based on the city-wide assessment and the household surveys. Working groups were constituted and led by the city government with strong stakeholder representation to increase people's resilience to climate change. The working groups address: (a) improving settlements and basic infrastructure; (b) enhancing livelihoods; and (c) developing climate and disaster risk management systems. Action area (d) improving environmental management and climate change mitigation actions, relates to the expressed commitment of the local government and the citizens to contribute their share to reducing GHGs while improving the local environment. Energy-saving measures are proposed for the operation of buildings and in the transport sector and two-stroke engines of local three-wheelers are to be converted to four-stroke engines.

A wide range of possible interventions have emerged from the working groups to make Sorsogon and its citizens more climate-resilient. They include: expanding the existing micro-finance scheme which targets in particular women, to include a micro-insurance for livelihood and housing, and the planting of local nut trees as wind breakers to reduce erosion and to provide additional income. Priority issues relate to shelter and disaster risk reduction. Based on their capacity and the need to diversify their livelihoods, selected community members will be trained in making houses in low-income communities more climate resilient and to help rebuilding these houses more quickly should they be destroyed. To further support this initiative, the city will stock tools and building materials.

Some women were complaining that schools were closed for too long after disasters as they functioned as shelters. The city is now committed to ensuring that schools would be retrofitted ensuring that studies can recommence shortly after disasters. This would help women to attend to the immediate needs after the disaster while their children would be in school. It was also decided that some schools would have to be relocated as they were originally erected in flood-prone areas and directly on the beachfront.

Concluding remarks

The initial participatory process led inadvertently to a more strongly developed gender assessment, addressing particular gender concerns and providing the city and community with some disaggregated data upon which more informed decisions can be taken in order to respond to climate change.

All actions were recommended by mixed-gender multi-sectoral groups and many of these recommendations clearly attempt to reverse the expected over proportional burden that disasters and by extension climate change would have on women.

However, much of the household level assessment only partly touched upon the prevailing gender division of labour. The assessment only provided a glimpse on the different risk perceptions of women and men.

Existing coping strategies were assessed, but a stronger focus on the different approaches of women and men may further benefit the development of climate change responses.

Gender roles in general and in particular in regard to possible roles of women to contribute to and to benefit from or suffer under adaptation and mitigation measures were not assessed in detail. Therefore the emerging climate change responses would benefit from a deeper gender assessment.

Reference

UN-HABITAT (United Nations Human Settlements Programme) (2009) 'Climate change vulnerability and adaptation assessment report', UN-HABITAT, Nairobi, prepared by Adelaida Mias-Mamonong and Reinero M. Flores

Part II
Realities on the Ground

5

Learning from Practice:
Case Studies

RESPONDING TO CLIMATE CHANGE IN VIETNAM: OPPORTUNITIES FOR IMPROVING GENDER EQUALITY

Koos Neefjes and Valerie Nelson

This case study reflects the outcomes of a study on gender and climate change executed in Vietnam (UN and Oxfam, 2009).

Introduction

Vietnam is particularly affected by climate change, despite good capacities in for example disaster risk reduction and agricultural research and development (R&D) when compared to other developing countries. Climate change responses can be equitable if they place women's empowerment centre-stage in climate change adaptation actions to increase resilience to climate change shocks and stresses. Actions to mitigate greenhouse gas (GHG) emissions and develop a low carbon economy must also be gender sensitive for achieving multiple goals. Climate change responses can create opportunities for achieving social goals but inequalities may be made worse without gender analysis and gender sensitivity.

Vietnam signed up to international agreements on gender equality and climate change, has national policies on energy efficiency and disaster risk reduction and is already implementing adaptation and GHG mitigation programmes. But there is a lack of policy coherence on gender and climate change. The 2008 *National Target Programme to Respond to Climate Change* (NTP-RCC) emphasizes gender equality as a guiding principle, but has no specific targets or activities to address women's vulnerability or gender issues in GHG mitigation. At the same time, climate change is not addressed in gender policy frameworks, such as the law on gender equality.

There have been some improvements over the past few years in capacity to deal with climatic stresses and to achieve greater gender equality, but there are still major gender gaps in actions to address climate resilience. Awareness of gender equality has increased in Vietnam, but this does not always translate into increased gender equality in practice, especially for ethnic minority women. There is also rising social inequality between rural and urban areas in Vietnam. Women's participation in household and community decision-making has increased, although generally less so amongst the poorest. But men continue to make the final decisions in relation to large expenses, and women's participation in local political and management structures remains low.

Gender aspects of climate change

Many Vietnamese are vulnerable to climate change, especially the poor in coastal areas, upland ethnic minorities and migrants in poor urban neighbourhoods. This is because they lack resources, are exposed to climate risks, face discrimination and depend on climate-sensitive livelihood activities. They are particularly vulnerable to water-related shocks and stresses that are getting worse because of climate change and affect different parts of Vietnam, such as typhoons, landslides, river floods and droughts, and sea level rise with associated saline water intrusion.

But climate change affects men and women differently. For example, there are disproportionate effects on women's mental health related to natural disasters, women often eat less in times of food shortage and suffer more health problems due to lack of clean water. Both men and women experience increased workloads during disasters, but women are primarily responsible for food supply and water collection, which is arduous; in fact they carry most of the burden.

This was shown for example in a researched community in the Quang Tri province in the central coast region of Vietnam, where typhoons are frequent. Men take the lead in disaster risk management (DRM) activities, such as rapid response teams, search and rescue, and protecting crops and livestock. Women focus on domestic responsibilities such as communal cooking, household food supply and caring for the sick and elderly, but also often take part in protection of crops and livestock.

Many women in rural areas are taking on more agricultural tasks as a result of male out-migration and local non-farm employment. They often replant paddy and plant subsidiary crops to supplement losses after disasters whilst they also must cope with less spectacular but unmistaken climatic stresses on crops. Women manage irrigation on their fields, yet men typically control the overall irrigation systems. Women also have limited access to other livelihood assets which would enable them to cope with major shocks and stresses. They have generally good access to micro-credit, with support from the Women's Union, for example. However, their access to larger scale credit is limited as women are less likely to have their names on land use certificates which are required as collateral for larger loans.

Women have important roles and responsibilities in the critical areas of disaster risk reduction, agriculture and household food security within vulnerable rural communities, but they are not yet fully involved in critical decisions to improve the resilience of households, communities, and regions. Better off social groups and richer communities adapt quickly to climatic and other shocks and stresses, they are more resilient. Women in such (rural and urban) communities tend to be better educated than their poorer peers and participate more in community affairs.

While education is seen as an escape route from poverty, girls leave school earlier than boys, and female illiteracy rates remain high in ethnic minority communities. Young people who do gain an education tend not to return to their remote communities after completing their studies. Out-migration includes young women employed in the garments industry who may remit funds to their family in rural areas, but salaries in such industries are low.

New GHG mitigation opportunities may strengthen livelihoods of poor rural women and men and increase their resilience for shocks and stresses. However, better off social groups are more able to benefit from larger scale technological opportunities or improved financial and social services, because they enjoy better education and access to markets. Many of the current opportunities are urban and industrial focused initiatives, in which large businesses play a key role – in which women in senior positions are under-represented.

Climate change adaptation and improving resilience

Climate change and also gender equality are covered regularly by the media, which is key to raising awareness. People are thus broadly aware of the impacts, but this knowledge of climate change and its impacts is still limited. The focus of action is mostly based on past experience with natural disasters and the capacities developed to respond to those. Most people lack resources to respond comprehensively and post-disaster coping and recovery activities tend to focus on restoring existing infrastructure and livelihood systems, rather than more transformative change which could increase household and community resilience as the climate changes. Gender analysis, comprehensive capacity building and community participation in planning and implementing disaster risk reduction measures is only happening in a limited number of localities, so far. This is about to change though, as a nationwide programme to scale up experiences with community-based disaster risk management (CBDRM) is now being prepared, which should mainstream both gender and climate change resilience.

In three rural communities, in-depth dialogues were held with local women and men and their (male and female) leaders at different levels on climate change vulnerabilities and existing capacities as well as current and future adaptation actions:

• Hai Ba and neighbouring rural communities in the central coast region are severely affected by climate change, such as increasingly severe typhoons and sea level rise. There is substantial livelihood diversity and household resilience, for example in flood-prone parts where mezzanine levels in houses are constructed for food and asset storage. However, several households will have to relocate their homes and livelihood activities to more secure areas. This will be doable for better-off households but others will need support as this is costly, whilst movement will affect many in the 'receiving' parts.

- Avao is a community in the uplands of the centre of Vietnam, populated mostly by ethnic minorities with low incomes. Shifting cultivation cycles are shortening due to population pressure and land fertility is being undermined, leading to increased soil erosion and heightened risks of landslides following heavy rains. More frequent forest fires are also likely due to prolonged droughts expected from climate change. A potential alternative to rice farming is planting trees for use in the pulp and paper industries, but villagers lack capital and knowledge of markets. Adaptation to climate change is limited by short-term needs in these poor communities.
- Dai Nghia is a rural community in the northern part of Vietnam, affected by the urbanization and industrialization of Hanoi and neighbouring provinces. Locals have observed unusual climate events (hailstorms, floods, longer periods of very cold or hot temperatures), but have limited experience of responding to major disasters. Many households, especially women, are still involved in agriculture and feel that it is unclear what they should do if faced with higher frequency and intensity of climatic stresses. They articulated no clear strategies for change and they may need outside support to adapt.

According to these dialogues in different parts of Vietnam, men and women agree on many measures in response to disasters and other climatic stresses. For example, the importance of adaptation measures such as shifts in cropping calendars (of rice) in relation to the typhoon and flood season were discussed, as well as the importance of drought or flood resilient crop varieties. There are also differences in what women and men identify as priorities. For example, women put greater stress on the importance of health care, food security and access to clean water. On the other hand, women usually do not actively take part in decision-making on natural resources management and disaster management. In addition, gender bias in agricultural extension was discussed in all these communities, as these services remain targeted to men whilst women now play a dominant role in small scale agriculture.

Only a few better off households can diversify their income sources, build houses which are stronger and more flood resistant, and/or have sufficient resources to relocate completely. Climate proofing of schools and health centres is not yet happening but will be increasingly important with benefits for poor women and girls from continued access to such facilities during and after climatic disasters and stresses, as they face the most pressures.

Young people in Vietnam are quite aware of disaster preparedness and have a reasonable understanding of broader environmental issues. In the study, teenage boys and girls suggested climate change responses, such as learning and investment in new technologies, investment in education, and awareness raising. Girls focused on small-scale technologies and social investment as responses to climate change and identified that they would like domestic responsibilities to be more evenly shared. Boys focused more on technological and industrial shifts when envisaging future responses to climate change.

Rural GHG mitigation, resilience and gender

Some GHG mitigation strategies can have benefits for rural poverty reduction and development, including payments for carbon sequestration to local forest managers under the Reduced Emissions from Deforestation and Forest Degradation (REDD) mechanism; the production of biogas from animal waste; and the System of Rice Intensification (SRI). There are substantial experiences with these options in Vietnam.

Vietnam is preparing itself for the forthcoming international agreement on a REDD mechanism, based on experience with payments for environmental services about which legislation is being prepared. The promise of REDD is that additional payments can become a major incentive for forest protection and maintaining the multiple environmental and social-economic roles that forests play, especially in ethnic minority communities in uplands and also in Vietnam's coastal belt where mangrove restoration, protection and expansion is needed. A focus on poverty reduction is critical, as globally many rural households depend upon forest resources for their livelihoods, and more than 70 per cent of people living in poverty are women. Also, women tend to have less secure forest tenure so REDD must support community forestry and build in guarantees for benefits to ethnic minorities and women's participation in forestry management.

Studies in Vietnam have identified various positive benefits of the production and use of biogas from animal manure, especially for women. Several projects have supported the construction and operation of tens of thousands of household level micro-digesters that produce methane especially for cooking. One project is aiming to bring the total in the near future to more than 200,000 across the country. Women's exposure to wood smoke and time spent collecting fuel and cooking has reportedly been reduced. Costs and reduced demand for fossil fuels has also occurred and other benefits include better hygiene in the yard, production of odourless fertilizer for fields, and increased employment opportunities (mainly for male workers in construction of the digesters).

The SRI is being implemented in many provinces of Vietnam and offers potential benefits by making crops more resistant to drought and storm damage while it helps reduce methane emissions (a strong GHG) from reduced water use and improved land tillage practices. Women's workloads in some of these activities increase but in others decrease, while income increases. Various analyses lead to the conclusion that with targeted efforts (for example, in terms of extension) gender equality can improve as a result of SRI, along with economic and environmental benefits. However, as with other technological innovations, women's strategic interests must be closely monitored in order not to be overlooked. Much analysis is needed of potential positive and negative impacts of new technologies on gender relations to ensure women can realize the potential benefits and that gender inequalities are not exacerbated by the introduction of new technologies.

Conclusions and lessons learned

Climate change affects poor people and especially poor women in a wide range of communities. Without gender analysis and gender sensitive action, climate change will set back human development, including gender equality. Promoting the empowerment of women and marginalized groups have to be central elements of an equitable response to climate change – it is not an optional extra.

Substantial social development experience is available in Vietnam and elsewhere for supporting gender sensitive climate activities and for promoting women's empowerment. If these experiences are integrated in responses to climate change then they become opportunities for improved resilience as well as greater gender equality. This gender mainstreaming in climate change action must happen in (scaling up of) disaster risk reduction activities; livelihood support activities including agricultural extension, forestry management and financial services; migration and resettlement policies; construction of schools and hospitals, to ensure accessibility during periods of stress; and in climate proofing of infrastructure such as irrigation systems, dykes, roads and bridges.

Techniques for mitigation of GHG emissions and sequestration of carbon can support livelihoods and health and thus enhance resilience for many kinds of shocks – they are therefore also adaptation measures. They also show that there is real, concrete potential for achieving the multiple goals of responding to climate change and improving gender equality, if they are gender-sensitive.

It is important that awareness on both climate change and gender equality is raised, with specific attention paid to curricula on both gender equality and climate change action and an improved research base on the gender and climate change links. It is also important that development programmes promote socio-ecological resilience in economic development rather than damaging ecosystems and livelihood resources.

Policies, programmes and projects to help improve women's participation in decision-making on climate change are critical and these initiatives should draw on the skills and agency of women and marginalized groups.

Reference

UN and Oxfam (2009) *Responding to climate change in Vietnam: opportunities for improving gender equality: A policy discussion paper,* UN and Oxfam in Vietnam, Hanoi, www.un.org.vn/index.php?option=com_docman&task=doc_details&gid=110&Itemid=266&lang=en, accessed 22 February 2010

Contributors to this discussion paper included Koos Neefjes, Vu Minh Hai, Valerie Nelson, Irene Dankelman, Tran Van Anh, Ingrid Fitzgerald, Bui Viet Hien, Julie Theroux-Seguin, Nguyen Nhu Hue, Nguyen Thanh Ha, Pham Thanh Hoai, Nguyen Thi Hoang Yen and Vu Phuong Ly.

CASE STUDY 5.2

GENDER DIMENSIONS, CLIMATE CHANGE AND FOOD SECURITY OF FARMERS IN ANDHRA PRADESH, INDIA[1]

Sibyl Nelson and Yianna Lambrou

Introduction

'There is dry drought followed by green drought.' This is how a farmer in Ankilla village of Mahabubnagar district, Andhra Pradesh[2] in the south of India, described the effect of the delayed monsoon on the growing season. When the late rains finally arrived, they only helped the weeds to grow and the farmers lost that year's crops. In a drought year such as this one, what do the farmers do when the rains come too late? What choices will they make to ensure that their families eat?

The cycle of drought is a recurring challenge in parts of Andhra Pradesh, and food security and income are tied to the variations of the weather (Acosta-Michlik et al, 2005; World Bank, 2006). This cycle appears to be increasing and becoming a chronic reality. Women, the caretakers of the family health and food security, are under increasing burdens to ensure food security while men are pressured to obtain additional income from non-agricultural sources (Lambrou and Nelson, 2010).

The potential impacts of long-term climate change in India could be severe, particularly in the agriculture sector (Brenkert and Malone, 2005; Mall et al, 2006). Increasingly unpredictable climate shifts are projected to increase the frequency and intensity of droughts in Andhra Pradesh (World Bank, 2006; Prabhakar and Shaw, 2008). At the same time, gender-based discrimination is still a reality in India and in Andhra Pradesh (Government of Andhra Pradesh, 2008) even though advancements have been made in social equality and development (APMSS, 2007). A growing body of research suggests that climate, development and gender challenges go hand-in-hand and some efforts have been made to link them (Government of

India and UNDP, 2008). More work is needed so that long-term institutional approaches for climate change adaptation are gender-sensitive and locally appropriate (Ahmed and Fajber, 2009; Kelkar, 2009).

This case is based on a study that sought to understand how farmers coped in the past with drought, what hardships they face now compared to 30 years ago and how gender roles determine what male and female farmers can and will do to ensure food security. Such information can provide crucial advice to policy-makers for building resilience in responding to future shocks and for adaptation to long-term climate change.

Climate change impacts through a gender lens

The research examined climate trends over time and specific extreme events (particularly droughts), and within both timeframes examined the gender aspects of livelihoods and food security at the individual and household levels. The following three gender specific aspects were examined:

- Division of labour: What do men and women do differently to ensure income and food security for the household in response to different climatic conditions in agriculture?
- Decision-making: How are decisions about household livelihoods and food security shared especially in response to the impacts of climate change?
- Access to information: Do women and men have equitable access to information on farming practices and weather?

Location and context

Smallholder farmers (with less than five acres of land) whose livelihoods depend primarily on rainfed agriculture in six villages of two drought-prone districts of Andhra Pradesh (Anantapur and Mahabubnagar) shared with us their views about their livelihoods and food security. Although the farmers will be referred to as 'farmers' here, farming was no longer their prime activity as for some time they have been seeking paid employment (either through day labour or longer-term migration) as an important source of income for meeting household needs.

In addition, the farmers' food security was not ensured solely through their farming activities. The study participants reported multiple food sources: government rations (95 per cent of study participants), crop production (85.1 per cent), purchase (84.6 per cent), livestock (8.5 per cent) and wild food (0.5 per cent). The links between climate change and the food security of these farmers is thus not only that the changing rainfall and temperature patterns affect the yield of the farmers' crops and in turn affect how much food they have to eat. The farmers' food security is also linked to the income they derive from employment in climate-sensitive sectors, water availability to prepare the food, and other factors. This study focused on some of the main livelihood activities of the farmers: changes in the agricultural production activities and off farm wage labour and migration.

Methodology

The hypothesis we wanted to explore was whether due to gender roles – the behaviours, tasks and responsibilities a society defines as 'male' or 'female' – men and women are diversely affected by and cope differently with climate variability and longer term change. We wanted to understand how long-term climate change in these drought prone areas would affect the farmers, whether men and women would experience climate change differently and in what ways they would respond.

The research began with the farmers themselves. A field team made up of professionals from a women's empowerment organization based in Andhra Pradesh conducted focus group discussions with separate groups of men and women. The initial exercises were framed as a discussion about men's and women's perceptions of the availability of water resources, major drought events and seasonal rainfall over the past 30 years and related farming activities and livelihood opportunities, including migration ('climate change' was not explicitly mentioned by the facilitators). When it was established from what the farmers' reported that there had been a change in temperature and rainfall between now and 30 years ago and that droughts are a recurring problem for the farmers, additional focus group discussions were held to probe deeper into vulnerabilities to extreme events (droughts), coping strategies, perception of risks, food security and institutional support. In addition, representatives of the men's and women's groups met together in order to cross-reference their observations.

Based on what the farmers reported, a survey was developed and implemented to quantify the trends the farmers had observed. The survey captured the views of 106 women and 95 men. In addition, climate analysis (using 30 years of temperature and rainfall data) and institutional analysis (based on key informant interviews) were undertaken to triangulate the results of the focus group discussions and the quantitative survey of farmers. The farmers' perceptions were substantiated by the climate analysis of the last 30 years.

Findings

The study found that men and women farmers are facing multiple challenges, including deforestation, indebtedness and chronic food insecurity. As has been documented elsewhere, farmers' livelihoods are no longer based solely on agriculture, and migration for wage labour is an increasingly important strategy. In this context, the study found that gender does make a difference in dealing with climate change.

Long-term climate change

Gender-based division of labour is reflected in men's and women's perception of climate change impacts. While both men and women farmers perceived that temperatures had increased over the past 30 years and that the timing of rainfall had shifted to later in the season, there were similarities and differences in how men and women described the impacts of these changes. Almost 90 per cent of men and women reported that climate changes had led to poorer harvests or reduced crop yields. Men were significantly more likely than women to report there was less fodder and that bore wells and ponds had dried up. Women were significantly more likely than men to report that health was affected. These perceptions were indicative of men's and women's roles in the household, and may indicate a difference between men and women in terms of prioritizing what aspects of household life need to be supported in a changing climate.

Gender-based preferences for livelihood strategies in response to long-term change have implications for adaptation decisions. When asked what they would do if there was not enough rain for a few years in a row, both men and women ranked taking wage labour as the preferred response. However, when asked what livelihood strategies they would adopt if the weather became unpredictable from year to year, men and women had markedly different preferences. Women prefer wage labour close to home (57 per cent of women, 38 per cent of men) while men prefer to migrate further away (47 per cent of men, 18 per cent of women). The difference in preference of coping with uncertainty between men and women has large implications for the future of communities and the location where people seek work, as well as the institutional support for them.

Food security is less linked to climate variability over time, but appears less nutritious for all. Farmers in the study area get the majority of their food from government rations and for 20 years at least have not been self-reliant in terms of their food sources. In comparing the food they eat now to 30 years ago, farmers reported that food was sufficient, but not as nutritious as it used to be as they were no longer eating their traditional crops. This trend may be less linked to a changing climate and more to the introduction of government rations recently, and the need to supplement them with cash-purchased food. Being less reliant on the vagaries of climate variability may be a positive trend for farmers; however it does not appear that their food security needs were entirely met.

Extreme climate events – drought

Decision making on a seasonal timeframe is based on gender roles. The current context of decision making at the farm level on what crops to plant is dominated by the male head of household, determined by rainfall, and not based on advice from the agricultural department. In the study sample, 60 per cent of respondents reported that they follow the husbands' decision, 28 per cent follow a joint decision and 7 per cent follow the female (no statistically significant data on female heads of household). Almost all farmers (96 per cent) said that rainfall is a factor that influences cropping patterns; the next most common factor (30 per cent of respondents) is investments. Farmers reported that they do not change their crop decisions on advice from the agricultural department. During a crisis such as a drought, however, decision-making tended to become a joint process. This suggests that droughts could upset the typical gender roles in decision making and perhaps allow for new social patterns and diverse roles and expectations to emerge.

Gender discrimination in access to information and institutional support affects women's participation in decision making. As expected, access to information among farmers overall was low, and it was lower among women than men. Here, 47 per cent of men compared to 21 per cent of women had access to information on cropping patterns. In addition, discrimination in institutional support was more likely to occur on the basis of gender rather than caste or land holding. Local institutions advised farmers on strategies for coping with climate but in the current context men were more likely to receive information. Women were often not aware of the available information nor had access to it, thus remaining even further removed from the decision making plans on responding to climate shocks.

Some gender-based traditional coping strategies are exacerbated by drought; others are no longer feasible. It appears that during drought the traditional coping strategy of women eating less was exacerbated (to cope with food scarcity during a drought year, 5 per cent of men noted that the women ate less, while 17 per cent of women reported that women ate less as a coping strategy). Longer term adaptation strategies would need to ensure that this coping strategy was not employed. In addition, it may be possible to draw on past strategies for coping with drought for positive outcomes, however some strategies, such as eating wild foods, are no longer possible due to deforestation or the increased use of so-called unused land for biofuel feedstock production.

Lessons learned, emerging trends

Gender does make a difference in dealing with climate shifts and their impacts on food security, as this research in the south of India showed. Both for the farmers responding on a daily basis and to policy-makers providing long-term institutional support, it was clear that development needs, including gender equality, cannot be separated from building resilience to climate change, but are not as yet addressed in unison.

Interventions must consider climate shifts in terms of what it means to the world's most vulnerable people, especially poor women and men farmers. The differences between men's and women's access to resources, as well as gender differences in selecting coping strategies, are beginning to be understood (Ahmed and Fajber, 2009; Kelkar, 2009). It is becoming imperative that they be addressed not only to ensure that women's contribution and well-being is assured, but that development goals are also realised (such as the Millennium Development Goals).

Planning for adaptation to long-term change must be founded on male and female farmers' knowledge and experiences as they make choices in an uncertain climate. In addition, future plans and government policies must consider the reality of farming today, including migration and what the implications will be for longer term community and family stability. Additional research is needed on the differences between short-term coping strategies and long-term adaptation. In pursuing strategies for the future, the needs of female and male farmers need to be incorporated in regional, national and international development plans to ensure an integrated and gender sensitive approach based on sound knowledge and strategies that build the resilience of the most vulnerable to the impacts of climate change.

Notes

1 This case study is based on a project carried out by the Food and Agriculture Organization of the United Nations (FAO) with funding from the Swedish International Development Agency (SIDA). Partners in the research included Samatha Gender Resource Centre and Acharya NG Raya Agricultural University (ANGRAU) and international consultants.
2 As of this writing, the government of Andhra Pradesh was addressing the move for the formation of a separate state of the Telangana region.

References

Acosta-Michlik, L., Galli, F., Klein, R.J.T., Campe, S., Kumar, K., Eierdanz, F., Alcamo, J., Krömker, D., Carius, A. and Tänzler, D. (2005) 'How Vulnerable is India to Climatic Stress? Measuring Vulnerability to Drought using the Security Diagram Concept'. Paper presented at *Human Security and Climate Change – An International Workshop*, Holmen Oslo, 21–23 June 2005, GECHS (Global Environmental Change and Human Security Programme), UNEP (United Nations Environment Programme), IHDP (International Human Dimensions Programme on Global Environmental Change), CICERO (Centre for International Environmental and Climate Research) and CSCW (Centre for the Study of Civil War), www.gechs.org/downloads/holmen/Acosta-Michlik_etal[1].pdf

Ahmed, S. and Fajber, E. (2009) 'Engendering adaptation to climate variability in Gujarat, India', *Gender & Development*, vol 17, no 1, pp33–50

APMSS (Andhra Pradesh Mahila Samatha Society) (2007) *Annual Report 2006–2007*, APMSS, Hyderabad

Brenkert, A. L., and Malone, E. L. (2005) 'Modeling Vulnerability and Resilience to Climate Change: A Case Study of India and Indian States', *Climatic Change*, vol 72, nos 1–2, pp57–102

Lambrou, Y. and Nelson, S. (2010) *Farmers in a Changing Climate: Does Gender Make a Difference? Food Security in Andhra Pradesh*, Food and Agriculture Organization of the UN (FAO), Rome

Government of Andhra Pradesh (2008) *Human Development Report 2007 – Andhra Pradesh*, Centre for Economic and Social Studies and Government of Andhra Pradesh, Hyderabad

Government of India and UNDP (United Nations Development Programme) (2008) *Women as Equal Partners: Gender Dimensions of Disaster Risk Management Programme, Compilation of Good Practices*, Government of India and UNDP, New Delhi

Kelkar, G. (2009) *Adivasi Women: Engaging with Climate Change*, International Fund for Agricultural Development (IFAD), The Christensen Fund, United Nations Development Fund for Women (UNIFEM), New Delhi

Mall, R. K., Singh, R., Gupta, A., Srinivasan, G. and Rathore, L. S. (2006) 'Impact of Climate Change on Indian Agriculture: A Review', *Climatic Change*, vol 78, nos 2–4, pp445–478

Prabhakar, S. V. and Shaw, R. (2008) 'Climate change adaptation implications for drought risk mitigation: A perspective for India', *Climatic Change*, vol 88, no 2, pp113–130

Rao, N. (2006) 'Land rights, gender equality and household food security: Exploring the conceptual links in the case of India', *Food Policy*, vol 31, no 2, pp180–193

World Bank (2006) *Overcoming Drought: Adaptation Strategies for Andhra Pradesh, India*, World Bank, Washington, DC

Case Study 5.3

The Gender Impact of Climate Change in Nigeria

Omoyemen Odigie-Emmanuel

This case study seeks to show convincingly that the nature of climate change impacts on men and women is mostly different in character and that planning and action for climate change – including adaptation and mitigation – must recognize, document and take into consideration these differences in order to develop policies, plans and strategies that are effective, efficient and bring about equality and equity in their development outcome.

Introduction

Nigeria has a total territorial area of 923,768km². Of this, its land mass is 910,768km², while its water area covers 13,000km² (Awosika et al, 1992). The 2009 estimate puts the population of Nigeria at more than 154 million. It is estimated that 64 per cent of this population are rural dwellers, of whom 45 per cent engage in agricultural activities such as fishing, farming and rearing animals (Ekong, 2003). The Nigerian coastal and marine area consists of a narrow coastal strip of land bordered by the Gulf of Guinea on the Central eastern Atlantic and stretches inland for a distance of about 15km in Lagos to about 150km in the Niger Delta and about 25km east of the Niger Delta. Fishing is a major activity in the coastal area (Folorunsho, 2009).

The major drivers of climate change in Nigeria are human and industrial activities. Human activities include such things as the felling of trees, the burning of charcoal and the expulsion of fumes from cars, generators and motorcycles. Also the import and use of industrial waste like second-hand cars contribute immensely to the production or emission of carbon dioxide (CO_2). The exploitation of oil and gas for export in the Niger Delta is a large contributor to climate change. Commenting on the problem of gas flaring by oil companies in Nigeria, Sharife noted that: 'The oil rich region ... possesses over 100 vertical and horizontal flares emitting over 45 billion kilowatts of heat. Many of these flares based at ground level are located close to the communities' (in Odock, 2009, p13).

According to the Nigerian Meteorological Agency (NIMET) meteorological data show that there is already clear evidence of climatic changes in Nigeria. This is seen in the shift of rain patterns, increase in the incidence of floods and erosion, temperature rise, dust haze occurrences and increases in coastal erosion. There is a shift in the onset and cessation dates of the rainy season across Nigeria. The rainy season has become shorter and the August break, which usually was a brief dry period in the rainy season, now occurs in July. Rain deficits resulting in dry spells occur in and around Zamfara, Adamawa, Kebbi and Kaduna states (Ekweozoh, 2009, p3).

The rate of desert encroachment is increasing. The desert area is estimated to have extended over 35 per cent of the land mass and is advancing at 0.7km per annum on average. The desert margin has moved from Latitude 12° 30 N (Kebbi, Kano, Maiduguri) to Latitude 10° 30N (New Bussa, Jos, Shellen), forcing the savannah – the interface region of the desert and the forest belt – to the region traversing Oyo, Osun, Kogi, Enugu, Ebonyi and Benue states. According to the Food and Agriculture Organization of the United Nations (FAO), the remaining forest area in Nigeria will likely disappear by 2020 if the current rate of forest depletion continues unabated (Ogbonnaya, 2003, pp2–3).

Climate change in Nigeria has exacerbated coastal area changes and has increased the incidences of erosion and flooding in the country. In May 1995 and July 2005, Victoria Island, Lagos, was flooded during an ocean storm surge. A community in Agulu-Nanka, in the eastern part of Nigeria, has been displaced by flooding and several people were killed.

Climate change impacts are also evident in reduced agricultural productivity arising from lower yields as a consequence of the changed weather pattern. Local farmers are no longer able to predict incidences of rains based on past observations and the impact of drought on the crops is frustrating their farming activities.

Heatwaves associated with an increase in temperature are on the increase in many parts of Nigeria, for example in the Zuba community of Niger State. Heatwaves have a huge negative impact on human health. According to *Global Warming and Health: Fast Facts*, a leaflet produced and distributed by the World Wide Fund for Nature (WWF) Climate Change Campaign: 'Due to warmer temperatures, the number of people at risk of malaria could increase to an estimated 60 per cent of the world's population. Global warming is projected to significantly increase the range conducive to the transmission of both dengue and yellow fever.' Climate change also results in negative impacts on mental health.

Climate change is also resulting in the drying up of surface water. Lake Chad, Nigeria's major lake has been reduced to about a quarter of its normal size. Clearly evident is the fact that the water area at Lokoja where the River Niger and Benue meet is greatly reduced from its normal size.

Climate change is exacerbating conflict in Nigeria. In an article entitled 'On the Threshold: Environmental Changes as Causes of Acute Conflict' (1991), Thomas Fraser Homer-Dixon listed Nigeria as one of the countries that is prone to environmentally induced national and international security challenges. The agitations in the Niger Delta region of Nigeria have become aggravated by the environmental challenges created by climate change and the related environmental pollution from oil exploitation, particularly the gas flaring activities of the oil companies (Agwu, 2009).

Gender impacts of climate change

The gender impact of climate change in Nigeria is rooted in the social and cultural roles that have been attributed to men and women over the years. Men are seen as heads of households and the bread-winners, they are involved in national politics, have access to lucrative careers and credit, and they control resources, while women are seen but barely heard and are the ones who farm, fish, produce, process and sell foodstuffs, care for livestock, collect water and firewood as well as bear and raise children, care for the elderly and sick and run the domestic household (Akinyode and Iyare, 2005).

The above listed social and cultural roles attributed to women and men not only determine their contribution and involvement in the sectors affected by climate change but also determine their understanding of the issue of climate change, the way they are impacted and their ability to adapt and mitigate the impact of climate change. It follows that discourse and action on climate change must take into consideration the gender impact of climate change.

Men and women are both involved in food crop cultivation and fishing. The impact of climate change on food productivity is already destroying and disrupting livelihoods. While men migrate from rural to urban areas to get jobs and often more easily find work as labourers in industries, women are stuck with the disruption of their livelihoods by changes in the climate. They cannot move as easily as men, as the care and nurture of the household and children is their responsibility.

Women in Nigeria who are responsible for securing food, water and energy for cooking will be disproportionately burdened by the gender implications of food insecurity and food shortages, because they have to work harder to secure these resources. That translates as more time and energy used for the provision of water, food production and fuel collection, and the need to resort to inferior sources of energy which increases their exposure to indoor air pollution.

When there is injury from flooding, more women are likely to die especially in the north because of a lack of preparedness and lack of access to information. Women have to give care and this will again increase the time spent in unpaid labour. Women have fewer assets to recover from natural disasters.

Temperature rise from climate change is leading to a rise in incidences of malaria and cerebrospinal meningitis, especially in the northern parts of Nigeria. Women bear the burden of taking care of the sick, mostly lacking access to health services and will be further burdened because loss of biodiversity will mean additional loss of their major medicinal sources of medicinal plants.

Climate change is already depriving people of the right to shelter. If the deteriorating weather pattern is not checked, floods are expected to be more severe. Experts have predicted that if the sea level rise is superimposed on the subsiding Niger Delta (exacerbated by oil and gas extraction), the net effect is that within the next two decades about a 40km-wide strip of the Niger Delta and its people would be submerged. The impact of this on women will be disproportionate because women in the region lack access to information and credit to enable them to prepare, and they do not own boats to enable them to escape at such times. Women also lack access to land and bear more of the workload in periods of reconstruction.

When there is a conflict, while more men will die, women are likely to suffer more from the indirect consequences; they bear the sole responsibility of running the home and are victims of rape and other abuses. Finally, women generally have less capacity to understand and participate in climate change discussions because they are predominantly illiterate, lack information and lack access to decision-making bodies.

Coping strategies: Climate change adaptation and mitigation in Nigeria

Recent activities in the Niger Delta show that women are becoming involved in mobilizing for social justice and are likely to be more involved in conflict resolution while protecting their environment and protesting against corporate abuses. This is reflected, for example, by incidences in the Niger Delta where women took over Chevron installations in Escravos, Abiteye, Makaraba, Otuana and Olera Creek (see Case Study 9.3). So far, though some men and women in Nigeria have noticed and are protesting about the disruptions in weather patterns and other impacts of climate change, few really understand the consequences and future dangers of climate change. In other words, few understand the need for preparedness and the role they need to play in climate change adaptation and mitigation.

Nigeria is a party to the United Nations Framework Convention on Climate Change (UNFCCC) and has prepared its first National Communication on Climate Change. Nigeria has also established a national focal point on climate change – the Special Climate Change Unit at the Federal Ministry of Environment. The National Assembly has also set up a Committee on Climate Change. Nigeria is in the process of commencing activities towards its National Adaptation Strategy and Action Plan. In this respect, a memorandum of understanding has been signed by the Special Climate Change Unit, United Nations Development Programme (UNDP), Building Nigeria's Response to Climate Change (BNRCC), Nigerian Environmental Study/Action Team (NEST) and Nigeria's Climate Action Network (CAN).

Some state governments are beginning to promote afforestation efforts in their states. Worthy of commendation is Lagos state government. Lagos state has responded by putting in place a special climate change unit and the government has set the pace by planting in the region of 1 million trees in two years (Attoh, 2009). The Lagos state government has also created incentives for people to reduce car emissions by providing Bus Rapid Transit (BRT) systems in their state. However, there is a huge gap in the Lagos state initiative because it does not integrate a gender perspective.

The government of the city of Katsina has also adopted an afforestation initiative in which women and youths are encouraged and paid to plant and nurture trees.

Civil society organizations in Nigeria are playing a critical role in climate change planning, advocacy and action. Amongst these groups are non-governmental organizations, such as NEST, Environmental Rights Action, Friends of the Earth Nigeria, Women's Environmental Programme, Center for Education and Leadership Development, the International Center for Energy, Environment and Development, the Community Research and Development Center, 21st Century and coalitions and networks such as the Coalition for Change and Nigeria Climate Action Network. Glaringly absent is an organization of gender advocacy on climate change.

Conclusions and lessons learned

Climate change impacts are already evident in Nigeria; and these impacts are different for men and women. Only by integrating a gender perspective can plans and actions become effective. Absence of a gender perspective at the planning level will lead to programmes that are ineffective, inefficient, widen gender gaps and lead to frustrations at the implementation level as budgets do not match with strategic needs and interests. It is therefore necessary to integrate a gender perspective in climate change planning and action at the national level, but also at the states and at the local government levels.

The government–civil society partnership for developing Nigeria's action plan on climate change is commendable. It is, however, necessary that such an initiative should not only recognize the need for a gender perspective, but should also have on board a team of gender experts to coordinate the activities.

It is necessary for other states to emulate Lagos state government's approaches and develop their own state action plans for climate change mitigation and adaptation which must integrate a gender perspective. The government of Nigeria should promote, create, subsidize and regulate the production of products that are low in carbon emissions, such as stoves for women. The Lagos state government initiative in providing BRT bus systems to reduce the number of private cars emitting CO_2 is commendable and should be duplicated by other states. This initiative should be reviewed to integrate a gender perspective.

It is necessary to educate women on climate change to enable them to effectively play their role as agents of change in mitigation and adaptation plans. Climate change responses and coping strategies should tap into women's indigenous knowledge; and women should also be part of traineeships in cases of technology transfer.

References

Agwu, F. A. (2009) 'The impact of climate change on Nigeria's foreign policy', presentation at the two-day International Conference on Climate Change and Human Security in Nigeria: Challenges and Prospects, Nigerian Institute of International Affairs, Nigeria

Akinyode, A. and Iyare, T. (2005) *The 11-Day Siege: Gains and Challenges of Women's NonViolent Struggles in Niger Delta*, Women's Advocate Research and Documentation Center, Nigeria

Attoh, F. (2009) 'Climate change and its impact on Nigeria's rural sociology', presentation at the two-day International Conference on Climate Change and Human Security in Nigeria: Challenges and Prospects, Nigerian Institute of International Affairs, Nigeria

Awosika, L. F., French, G. T., Nicholls, R. J. and Ibe, C. E. (1992) 'The impact of sea level rise on the coastline of Nigeria', in proceedings of IPCC Symposium on the Rising Challenges of the Sea, Margarita Island, Venezuela, 14–19 March 1992

Ekweozoh, P. (2009) 'Technology transfer issues in climate change', presentation at the two-day International Conference on Climate Change and Human Security in Nigeria: Challenges and Prospects, Nigerian Institute of International Affairs, Nigeria

Ekong, E. E. (2003) *An Introduction to Rural Sociology*, Dove Educational Publishers, Ugo, Nigeria

Folorunsho, R. (2009) 'Coastline erosion and implications for human and environmental security', presentation at the two-day International Conference on Climate Change and Human Security in Nigeria: Challenges and Prospects, Nigerian Institute of International Affairs, Nigeria

Fraser Homer-Dixon, T. (1991) 'On the threshold: Environmental changes as causes of acute conflict', *International Security Journal*, vol 16, no 2, pp76–116

Odock, C. N. (2009) 'The political economy of climate change in South-South, Nigeria', presentation at the two-day International Conference on Climate Change and Human Security in Nigeria: Challenges and Prospects, Nigerian Institute of International Affairs, Nigeria

Ogbonnaya, O. (2003) 'Deforestation in Nigeria: Consequences and solutions', *Nature Watch*; Nigeria's Environmental Magazine, December 2003, pp.6–7

Case Study 5.4

Gendered Vulnerability to Climate Change in Limpopo Province, South Africa

Katharine Vincent, Tracy Cull and Emma R. M. Archer[1]

It is well-known that women in developing countries tend to be more dependent on natural resources than men, and thus that female-headed households tend to rely more on agricultural livelihoods than male-headed households (Meinzen-Dick et al, 1997). Climate change is projected to impact such livelihoods by altering the availability and distribution of natural resources, reflecting changes in temperature and the quantity and distribution of rainfall. Nevertheless, vulnerability to climate change depends on more than just the nature of exposure to the different climate parameters: it also reflects entitlements to the assets required to respond to the variations in climate. These assets include health, governance and political rights, social capital and networking, as well as financial and physical resources, and access to them is socially differentiated along lines of gender (as well as other aspects of social identity, such as age, ethnicity, class and religion) (Adger and Vincent, 2005; Denton, 2002; Cutter, 1995). This case study illuminates the differential vulnerability to climate change between female and male-headed households in a rural dryland community in Limpopo province, South Africa.

South Africa presents a particularly interesting context in which to look at gender relations and how they play out at the household level. A new constitution was intensely debated and negotiated at the time of transition to democracy in 1994, and it is now widely considered to be one of the most progressive in the world. Based on the principles of democratic values, social justice and fundamental human rights, equality features prominently, and thus the equal position of men and women in society is enshrined at the highest level (Government of South Africa, 1996). That said, the constitution also protects the rights of traditional customary law (and its patriarchal nature) alongside the new democratic governance institutions and processes. The result is that household decisions are embedded in an increasingly plural institutional landscape where the reality is that access to entitlements still differs between men and women, giving rise to gendered vulnerabilities.

The research used multiple social science methods to explore how vulnerability to climate change is gendered, and the gendered nature of access to coping strategies and adaptation in the recent past, as well as investigating how recent political and institutional change has affected the vulnerability of households of different headship. Data were derived from participatory rural appraisal, a questionnaire incorporating livelihoods survey, semi-structured interviews with household heads and key informants, institutional analysis, and policy analysis (for more information, see Vincent 2007a).

A naturalistic, place-based enquiry was undertaken in one community in Limpopo province, northeast South Africa, chosen for its experience of recent past climate variability and projected exposure to future climate change. This area, to the north of the Soutpansberg mountains, is semi-arid with a summer rainfall season (November–March). There are high levels of inter-annual variability, punctuated by regular droughts and occasional floods, most notably in 2000. In terms of human characteristics, the village comprises approximately 700 people in 180 households, and has a legacy of natural resource-dependent livelihoods (primarily crop farming, but also some livestock). From 1979–1994 this area was part of the 'independent' homeland of Venda, but since 1994 and its reintegration into South Africa, and the new and diversified opportunities that this has brought, there has been a steady shift away from the land. However, South Africa's high levels of unemployment impinge on access to formal sector employment, as does the legacy of poor education (with low levels of competency in English).

Vulnerability is determined by a number of driving forces, and the outcome of vulnerability is thus the result of an interaction of these. An overall household level index of social vulnerability to climate change-induced changes in water availability was constructed, comprising the weighted aggregation of five component sub-indices which were selected and weighted as appropriate to the specific context: economic well-being and stability (20 per cent); demographic structure (20 per cent); interconnectivity in higher level processes (20 per cent); natural resource dependence (20 per cent) and housing quality (20 per cent) (Vincent, 2007b). These sub-indices were theoretically

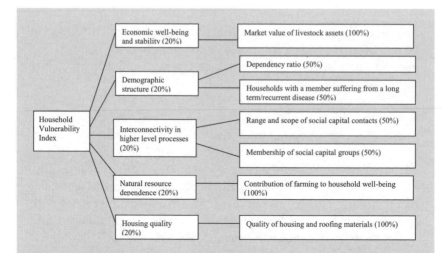

Figure 5.4.1 *Structure of the Household Vulnerability Index*

derived and based around the sustainable livelihoods framework (natural, social, human, financial and physical capital) (Adger and Vincent, 2005; Scoones, 1998) (see Figure 5.4.1). The overall index is normative, ranking households from the most vulnerable within the village, to the least vulnerable within the village. However, as the rank reflects an aggregation of five sub-indices, it is possible that a household may score poorly (i.e. be vulnerable) in four sub-indices, but a high score in the fifth sub-index could reduce its aggregate score such that in its overall rank it appears to be not so vulnerable. Table 5.4.1 summarizes the average vulnerability ranks of households of male-headed, *de facto* female-headed, *de jure* female-headed and child-headed households, where 1 is the most vulnerable.

Arguably the most striking observation from Table 5.4.1 is that there is no clear-cut relationship between household headship and level of vulnerability to climate change-induced variations in water availability. The average vulnerability ranks place *de jure* female-headed households as the most vulnerable, with an average of 40.27, but this is closely followed by male-headed households and then *de facto* female-headed households. Looking at the range of vulnerability ranks illuminates why these averages are so similar: the most vulnerable household in this village (with a rank of 1) is male-headed, and the second most vulnerable is *de jure* female-headed; but the range for both types of female-headed and male-headed households are large. Of particular note is the relatively low average rank of

Table 5.4.1 *Summary of average vulnerability ranks and range of vulnerability ranks by type of household headship*

Household type	No in sample	Average vulnerability rank	Range of vulnerability ranks
Child-headed	5	78.3	67.5–85
Male-headed	28	40.5	1–81
De facto female-headed	17	42.35	7–77
De jure female-headed	35	40.27	2–84

child-headed households, and their range of 67.5–85 (relatively less vulnerable). This is counter-intuitive and contradicts the literature, which suggests that child-headed households are typically vulnerable (Bicego et al, 2003). However, when looking past the end ranks to the composition and qualitative status of the households, each of the five child-headed households has this headship status not due to being orphaned, but because both their parents are working away in the city and have left their teenage children at home in the rural area to finish their schooling. Thus the child heads of these households are aged 16 and 17, and have good stores of financial capital through remittances from their parents, no dependence on natural resources (since their parents are employed in formal sector employment) and good levels of physical, human and social capital (through their interconnectivity and relationships with people outside of the immediate proximity of the village).

While the aggregate ranking is important to allow for the substitutability of strengths and weaknesses in the different capitals (and thus different sub-indices), it is at the higher level of resolution that the gender differences are most apparent. Table 5.4.2 shows that in the natural capital (dependence on natural resources) sub-index, *de jure* female-headed households had the highest level of dependence, making them more vulnerable to exposure to climate change. Here 79 per cent of male-headed households and 88 per cent of *de facto* female-headed households had only partial or no dependence, meaning that their livelihoods are not solely tied to the availability of natural resources, making them less vulnerable overall in this category. In contrast, Table 5.4.3 shows the ranks of the social capital (interconnectivity in higher level processes) sub-index, which itself comprises two indicators. While male-headed households tend to be less vulnerable on the basis of having a large number and wide range of contacts, *de facto* female-headed households are less vulnerable in terms of membership of local groups, such as savings wheels, burial societies and stokvels.[2] Women typically invest more, and gain more, from the reciprocity and networks of such local level social capital, which can provide both financial and psycho-social support in times of crisis (Goulden et al, 2009; Westermann et al, 2005). Many *de jure* female-headed households would also like to participate more in such groups, but are often constrained by the inability to pay the monthly membership fees. Box 5.4.1 provides a qualitative description of two of the sample households in order to further illuminate the profiles of vulnerability.

Table 5.4.2 *Natural capital sub-index scores disaggregated by household headship*

Household type	Frequency		
	Group 1 – heavy dependence	Group 2 – partial dependence	Group 3 – no dependence
Child-headed	0	0	5 (100%)
Male-headed	6 (21%)	8 (29%)	14 (50%)
De facto female-headed	2 (12%)	8 (47%)	7 (41%)
De jure female-headed	11 (31%)	8 (23%)	16 (46%)

Table 5.4.3 *Social capital sub-index scores disaggregated by household headship*

Household type	Mean indicator rank-contacts	Mean indicator rank-groups	Range of indicator ranks-contacts	Range of indicator ranks-groups
Child-headed	40.5	61.1	16.5–65.5	39.0–83.5
Male-headed	46.9	43.2	16.5–65.5	7.5–83.5
De facto female-headed	41.9	47.7	16.5–65.5	7.5–83.5
De jure female-headed	40.6	38.7	16.5–65.5	7.5–83.5

Box 5.4.1 *Profiles of vulnerability of two sample households*

The household ranked the tenth most vulnerable in this community is *de facto* female-headed. The female head is only 23 and she has only been head for a short period of time, since her husband left to seek work in Johannesburg. She receives irregular remittances from him, since he has not yet found permanent work, and has other financial assets in the form of a savings account and some small livestock. She cannot afford to belong to any social groups at the moment, and has medium dependence on natural resources as she farms maize for subsistence only.

The household ranked 75th (less vulnerable) is *de jure* female-headed. The female head is middle-aged (47) and has been head since the death of her husband seven years ago. She lives with her elderly mother, who receives a monthly social pension, and has a son working in Johannesburg as a policeman, who sends regular remittances. She is also fortunate to have formal sector employment as a saleslady. Her financial security allows her to belong to two social groups (a burial society and a stokvel), and she lives in a brick house. She does not farm at all, and thus has no dependence on natural resources.

Explaining the differences in ranks between households requires looking beyond the status of headship to the causes of that headship. The status of household headship is transient, with women in particular typically being part of male-headed, and *de facto* and *de jure* female-headed households at some point in their life. In particular in this community many of the *de jure* female heads of household were elderly, with an average age of 58.14 (Table 5.4.4), and had survived their husbands due to the gendered difference in life expectancy.

Table 5.4.4 *Relationship between age and household headship*

Household headship	Average age of head (years)	Range of ages of heads (years)
Child-headed	17.2	16–18
Male-headed	58.71	29–86
De facto female-headed	46.71	22–78
De jure female-headed	58.14	30–86

As well as differences in current vulnerability, access to coping strategies and adaptation shows gendered differences. Existing strategies in drylands to cope with inter-annual climate variability are based on flexibility in livelihoods, such as changing planting dates, planting hardier varieties (for example sorghum rather than maize), planting in alternative locations, or using river or borehole irrigation (Corbett, 1988; Ellis, 1998; Goulden et al, 2009). Men and women have different access to such options: land rights for women are often poor, and their lack of control of household financial capital and decision-making capacity may make it difficult to obtain new seed or sink boreholes for irrigation (Eriksen et al, 2005). Gendered roles that render women in charge of reproductive tasks, such as child rearing and healthcare, also mean that they have fewer options in terms of coping and adaptation, and thus tend to be more vulnerable than men. Men's gendered role is as the bread-winner within the household, and they have the ability to command other livelihoods as they can migrate, and have a historical legacy of better levels of education (although this is slowly changing), which can allow them to be more insulated from exposure to climate change. *De facto* female-headed households sometimes have the best of both worlds, as the household benefits from income from the non-resident male combined with insulation from the vagaries of climate, whilst the *de facto* female head often has greater decision-making powers in her husband's absence.

There are important implications of the empirical findings of the way in which vulnerability and access to coping strategies and adaptation within this community are gendered. Whilst many development policies within South Africa have directly or indirectly reduced absolute vulnerability to climate change, institutions and policies are rarely gender neutral, and thus the vulnerability of male and *de facto* female-headed households has, on the whole, reduced more than it has for *de jure* female-headed households. Gender differences in roles, responsibilities and capabilities mean that climate change may actually reinforce disparities between men and women. As a result, it is vital to consider the gendered effects of policies, to ensure that they do not inadvertently contribute to differences in the relative vulnerability to climate change of female- and male-headed households.

Notes

1 This chapter is based on primary research undertaken for Katharine Vincent's PhD thesis.

2   Stokvels are a South African example of a Rotating Savings and Credit
    Association (ROSCA).

References

Adger, W. N. and Vincent, K. (2005) 'Uncertainty in adaptive capacity',
    Comptes Rendus Geoscience, vol 337, pp399–411

Bicego, G., Rutstein, S. and Johnson, K. (2003) 'Dimensions of the emerging
    orphan crisis in sub-Saharan Africa', *Social Science and Medicine*, vol 56,
    no 6, pp1235–1247

Corbett, J. (1988) 'Famine and household coping strategies', *World
    Development*, vol 16, no 9, pp1099–1112

Cutter, S. L. (1995) 'The forgotten casualties: Women, children and
    environmental change', *Global Environmental Change*, vol 5, no 1,
    pp181–194

Denton, F. (2002) 'Climate change vulnerability, impacts and adaptation:
    Why does gender matter?', *Gender and Development*, vol 10, no 2, pp10–21

Ellis, F. (1998) 'Household strategies and rural livelihood diversification',
    Journal of Development Studies, vol 35, no 1, pp1–38

Eriksen, S. H., Brown, K. and Kelly, M. (2005) 'The dynamics of
    vulnerability: Locating coping strategies in Kenya and Tanzania', *The
    Geographical Journal*, vol 171, no 4, pp287–305

Goulden, M., Naess, L. O., Vincent, K. and Adger, W. N. (2009) 'Diversi-
    fication, networks and traditional resource management as adaptations
    to climate extremes in rural Africa: Opportunities and barriers', in
    W. N. Adger, I. Lorenzoni, and K. O'Brien (eds) *Adapting to Climate
    Change: Thresholds, Values and Governance*, Cambridge University Press,
    Cambridge, pp448–464

Government of South Africa (1996) 'Constitution of the Republic of South
    Africa, no 108 of 1996', Pretoria

Meinzen-Dick, R. S., Brown, L. R., Feldstein, S. L. and Quisumbing,
    A. R. (1997) 'Gender, property rights and natural resources', *World
    Development*, vol 25, no 8, pp1303–1315

Scoones, I. (1998) 'Sustainable rural livelihoods: A framework for analysis',
    IDS Working Paper no 72, Brighton

Vincent, K. (2007a) 'Gendered vulnerability to climate change in Limpopo
    province, South Africa', PhD thesis, University of East Anglia, Norwich

Vincent, K. (2007b) 'Uncertainty in adaptive capacity and the importance
    of scale', *Global Environmental Change*, vol 17, pp12–24

Westermann, O., Ashby, J. and Pretty, J. (2005) 'Gender and social capital:
    The importance of gender differences for the maturity and effectiveness
    of natural resource management groups', *World Development*, vol 33, no
    11, pp1783–1799

CASE STUDY 5.5

GENDER PERSPECTIVES IN ADAPTATION STRATEGIES: THE CASE OF PINTADAS SOLAR IN THE SEMI-ARID REGION OF BRAZIL

Thais Corral

For more than 50 years, the northeastern region of Brazil, where 45 per cent of the country's poor live, has been facing severe droughts. This situation is aggravated by climate change. The most vulnerable are women and children. Men migrate to the southern region to find jobs and often they do not come back.

In 2006, the Brazilian NGO REDEH (Network of Human Development) together with the SouthSouthNorth network started implementing an adaptation strategy focusing on irrigation for small agriculture as well as commercialization practices. The pilot project took place in Pintadas, a 11,000-people municipality, which has a long history of women's mobilization and leadership. The Pintadas Solar project has become a model and is being replicated in several other communities in the region around Pintadas. The initiative is now called Adapta Sertão. This case study presents some of the strategies used and the lessons learned. After two decades of work in the field of gender and environment, the experience has shown that addressing the climate crisis can be used to empower women and build the foundations for a new type of development that is gender sensitive and able to produce effective outcomes.

General context

The area where Adapta Sertão is currently operating is a semi-arid region that is among the poorest in Brazil. It is characterized by a climate with recurring drought periods and low and irregular precipitation (200–1000mm per year), affecting approximately 2 million families. Of all Brazilians living in poverty, approximately 45 per cent (18.8 million people) live in this region. Crop agriculture (cassava, maize, beans and green produce) together with cattle rearing is the primary economic sector of the region. Between

20–50 per cent of precipitation in the region occurs in downpours – intense and concentrated rainfall (from ten minutes to two hours in duration on average), that doesn't allow the water to infiltrate the soil properly. Since the 1970s, the Brazilian government, together with international organizations such as the World Bank, has been constructing hundreds of artificial reservoirs or earth dams in order to take advantage of the water from these downpours in this semi-arid region. The reservoirs were built to collect water for domestic use in the intense drought periods, cattle farming and crop agriculture. However, the latter aim of using water for agriculture in the semi-arid municipalities neglected the needs of small farmers, especially women and youths. During the pilot phase of the Pintadas Solar project and the installation of irrigation systems, the main barriers limiting the dissemination of efficient irrigation and production systems among small farmers that live in the area were indentified:

- **Limited access to technology and technical assistance:** Water pumps and drip irrigation systems designed for semi-arid regions exist but the commercial distribution network doesn't reach these municipalities. Therefore, the technology, when it does exist, is too expensive. There is also a lack of technical training and assistance to meet local needs.
- **Lack of low cost agricultural inputs:** Drought resistant seeds, soil improvers, the production of organic fertilizers and low-cost sustainable pest control methods are all fundamental tools to maximize effective agricultural production in semi-arid areas. However, it is rare that government projects support integrated production strategies using these essential tools.
- **Lack of finance and market access mechanisms:** The Brazilian government offered subsidized credit through the National Program for Family Farmers (PRONAF). However, unfortunately this was not utilized sufficiently to increase the agricultural production of the family farmer due to the lack of orientation and technical training given to the farmers and to the managers of the rural banks who continue to recommend and support inefficient conventional technologies. There is no specific training on gender either, even if women are the ones that stay in the semi-arid region when the men migrate. Moreover, effective strategies to increase market access and commercialization for the products produced by small farmers are essential to generate income.

Aside from these points, the current inefficient agricultural production systems in the semi-arid region have these main consequences:

- Continued poverty in the region.
- Women's marginalization from concrete social and economic opportunities.
- Malnutrition of children living in rural areas.
- The deforestation of the Caatinga, the local biotope.
- Rural migration to urban areas which is at the base of many social problems of most large urban areas (such as disintegration of the family unit, expansion of urban slums, social disparities; see Chapter 4); men migrate while women and children stay behind.

Situation of women

Generally speaking, the status of women in the semi-arid region is very low. Deprived from access to clean water and to the appropriate technologies, they are pushed to live in poverty. This means that they have very difficult lives in which very simple tasks such as cooking, washing and taking care of the household demand an enormous amount of work. In Pintadas, the situation is a bit different and that was one of the reasons for choosing to implement the project there. More than two decades ago, women started to be at the forefront of social mobilization in Pintadas. Organized with the help of the progressive Catholic church, a group of people, led by women, started a big struggle to possess land owned by few proprietors. The struggle known as Lameiro, which was the name of an extensive piece of land, was successful. The land could be occupied by hundreds of landless people. The Lameiro struggle gave women confidence and they made this event a landmark in the region. During the years that followed, a group of women formed the Association of Women of Pintadas and succeeded in electing in 1996 the very progressive woman mayor Neuza Cadore. This development was totally unusual in Sertão. Pintadas started a new system of politics that became known all over Brazil. The municipal administration won an award for its innovations in the public sector given by the prestigious Fundação Getulio Vargas University.

However, as in other places of the semi-arid region, women and youth in Pintadas are among the most affected by desertification and climate change. Within the family unit they are the ones that lag behind when the head of the family migrates to other regions of the country in the search for work.

Small-scale agriculture remains an opportunity for social and economic empowerment of women in the semi-arid region. However, socially, the man continues to be considered the head of the household, he is the one to whom credit and training are offered. Moreover, the lack of appropriate technology makes the work with land very time-consuming and ineffective, and adds to the burden of women or children that end up doing most of the everyday tasks.

The pilot project Pintadas Solar was designed to address this reality. Its main objective was to identify appropriate technology and a model for climate change adaptation that could also foster a new model of development.

The Pintadas Solar project

REDEH (Network of Human Development) and CEMINA (Communication, Education, Information), two Brazilian non-profit organizations, have worked with women in Pintadas since 1999. The focus of the collaboration was firstly the women's radio station of Pintadas. Radio Pintadas was very lively and stimulated new possibilities, one of them was access to the internet. In 2003, during a meeting with women to plan the new digital telecentre that was going to be installed, one of them raised a question: 'It is great to have access to computers and to the internet, but when are we going to get rid of the bucket?' She was referring to the manual way of watering the land.

All of those present at the meeting laughed. At that time the group did not imagine that three years later Pintadas would be hosting an irrigation project and the bucket would be replaced by very advanced small scale irrigation technologies.

The focus of Pintadas Solar was to test small scale irrigation technologies that were water and energy efficient and could ensure food security and income generation for farming families that otherwise would not be able to stay on their land. In two years, two different type of systems, drip irrigation and organoponia – a Brazilian technology that does not use chemical fertilizers – were tested, revolving funds were created and the local cooperatives were transformed in retailers of the irrigation systems, directly connected to the manufacturers. Pintadas Solar won the 2008 Seed Awards and became known in Brazil as a promising model to adapt local populations to climatic changes in the semi-arid region of Brazil.

Pintadas Solar, gender and climate change

In order to address gender and climate change, Pintadas Solar used two approaches.

At the operational level, Pintadas Solar identified a series of indicators that guided the project activities. The project team paid particular attention to the family as the unit. Both the woman and the man were treated on an equal basis to receive the systems, the training and the micro-credit. The training included participation in external events, travel and meetings. Whenever it was possible, women were chosen to participate.

The leadership of the project was also assigned to a woman farmer. She chose a team mainly formed by young people that already had a different mentality with regard to gender relations.

The second approach is associated with the feminine qualities that are imbedded in the project and the way it is conducted. Differently from many similar projects, the technical component, even though central, does not dominate the entire scene. Communication, cooperation, responsibility, respect and appreciation are cultivated. The beauty and care of the spaces in which the project activities take place is one of the characteristics of Pintadas Solar. Project participants appreciate this care and immediately feel more confident, which makes things much easier when challenges have to be faced to overcome the inevitable obstacles. There is not a project leader, but a leadership team. Every single stream of knowledge is validated so that all the participants feel they own the process and the project. In reality, participants are like partners of the project.

Adapta Sertão

In 2009, the phase of replication started and Pintadas Solar became a model now being implanted in four other municipalities. It is the main adaptation to climate change project of the semi-arid region of Brazil. Key innovations of Adapta Sertão are:

- **Supplying small farmers with technology and knowledge to make them resilient to climate change impacts.** The lack of access that small farmers have to technologies is overcome by creating a distribution network of efficient irrigation technologies based on local farmers cooperatives and organizations. These are not only 'distributors' and 'retailers' of technologies but also disseminators of modern agronomic knowledge to make crops more resilient to climate change. Training is a key component of the project. Women and youths are the most relevant groups to be taken on board.

- **Identification of new market niches linked to climate change that can add value to the production of small farmers.** Climate change represents an opportunity to bring a new perspective to old problems. The project is now developing a 'climate seal'. This will help with identifying crops and production processes that are 'climate proof'. Adapta Sertão is also getting involved in Fair Trade networks in Brazil. The 'climate seal' includes several indicators that are gender sensitive. The idea is to make sure that the effort to overcome the climate crisis cannot be dissociated of other very important social issues such as gender justice.

Current project activities are:

- Refining the distribution network model of irrigation technologies through local cooperatives in which the emphasis is to benefit women and youths; developing full capacity building modules on the distribution network, writing a business plan for scaling up; creating a structured partnership with national and/or international micro-credit institutions.
- Setting up an Adapta Sertão revolving fund in five additional municipalities of the Sertão to test the model in different contexts, build critical mass and allow small farmers to purchase irrigation technologies. Building the capacity of staff so that they incorporate gender sensitiveness among other things.
- Testing the refined arrangement (distribution network and revolving fund) in these five new municipalities allowing access to irrigation technologies to a minimum of 100 farmers within two years; of these, 50 per cent should be women.
- Developing a 'climate seal' as a voluntary certification for food products that make small farmers more resilient to climate change in the context of gender equality.

- Identifying of products that could receive the 'climate seal' and that could be sold in regional, national and international Fair Trade markets.
- Creating of a social enterprise that can disseminate the model in Brazil and worldwide.
- Organizing a series of seminars and workshops for the dissemination of lessons learned and to foster partnership building.

Looking forward

The urgency imposed by the climate crisis offers an opportunity to revisit development strategies and incorporate many of the analysis and recommendations that the women's global environmental movement made since the UN Conference on Environment and Development of 1992. For this to happen we need more women or men sharing the right perspective to occupy leadership positions. We see these changes happening in small projects that are mushrooming around the world. Yet, they are still not able to change the course of events.

The climate crisis urges the need for a paradigm shift. As Einstein stated, a problem cannot be solved with the same level of awareness that created it. What is the knowledge that we will use to build this new level of awareness? How can women contribute? Values such as care, collaboration, respect for nature and for other species, that for so long the women's movement has been an advocate for, are becoming an imperative for our survival. Experiences such as Pintadas Solar offer a ground for implementing this vision, that hopefully can inspire many other communities.

References

Adapta Sertão, www.adaptasertao.net
Pintadas Solar, www.pintadas-solar.org

CASE STUDY 5.6

CLIMATE CHANGE AND INDIGENOUS WOMEN IN COLOMBIA

Marcela Tovar-Restrepo

Introduction

Use, control and ownership of and access to natural resources by indigenous peoples is crucial for their cultural survival and sustainable development. Indigenous traditional territories provide them with their livelihood and guarantee their cultural integrity. This is especially significant for indigenous women who greatly depend on their environment for their family's survival, the maintenance of their cultural identity and cosmogony, their medical practices, the exchange of seeds, plants or other food sources, and their political agency and participation both within their own peoples and outside. This case study aims to illustrate the disproportionate impact of climate change on indigenous women's livelihood and cultural survival in Colombia. It also presents coping strategies, adaptation abilities and mitigation measures used by indigenous women to overcome these impacts.

The total population of Colombia is 41,468,384 people, of which 1,392,623 (3.43 per cent) are considered to be indigenous. Of that, men constitute 50.4 per cent and women constitute 49.6 per cent (DANE, 2009). The majority of the indigenous peoples reside in rural regions within 254,879km^2 of Colombia, which is 1,142,142km^2 total. There are 87 different indigenous cultures/peoples and 64 different languages. These peoples inhabit a wide variety of Colombian ecosystems: tropical rain forests and jungles, high mountains (some settlements in the Andes are as high as 5700m above sea level), valleys, forests, coasts (of both the Atlantic and the Pacific Oceans) and deserts.

The livelihood and the self-subsistence economies of the indigenous peoples are centred on agriculture, fishery, pasturage, hunting, gathering and horticulture in rural areas. However, for a great proportion of indigenous women, mobility and access to urban environments is essential for their economies. They need to exchange, buy and sell products important for their subsistence in urban markets. Because of the gender division of labour in most of these cultures, women work in the domestic/reproductive sphere, which involves food production in their small house plots (*chagras*), and food distribution and food conservation within their domestic units and their communities. Despite the fact that most commonly indigenous women do not hold decision-making positions and are not regarded as traditional authorities among their people, they are unique bearers of traditional knowledge systems and cultural practices and abilities employed to maintain biodiversity and environmental sustainability. Women's transmission of cultural knowledge to young boys and girls regarding access to and use of natural resources and preservation of biodiversity, guarantees the spiritual life of their peoples, the performance of their rituals and their traditional medical practices.

Ethnic and gender aspects of climate change impacts

Because of their gender and ethnicity, climate change consequences affect indigenous women more strongly, deepening exclusionary and discriminatory practices already present within their own peoples and in the non-indigenous majority of the society. Different conditions and circumstances make indigenous women a highly vulnerable population. In the first place there is their geographical location, generally indigenous women live in rural, remote areas where access and mobility are difficult. Second, women depend directly on their environmental livelihood as the only source for their subsistence. Third, illiteracy and poor knowledge of Spanish (the official language) prevents them from getting adequate information on climate change consequence mitigation and disaster prevention measures. Fourth, indigenous women suffer disproportionate displacement caused by internal armed conflict. Indigenous women constitute 38 per cent of the total displaced population in Colombia, which is estimated at 4 million people (CODHES, 2004, p14; Oxfam, 2009, p6). Most of them are forced to move into urban centres where they face obstacles in meeting their basic living needs. Fifth, their poverty and lack of monetary and non-monetary means to access goods and services in urban contexts or to participate in urban labour markets. Sixth, indigenous women are invisible as right claimants or as a social group with particular needs who should be the object of conflict mitigation policies and programmes. And finally, their own world-vision, cultural practices and social networks are needed to guarantee their agency and autonomy (ECLAC, 2006, p10; World Bank, 2004; UN Permanent Indigenous Forum, 2004).

Climate change in Colombia disproportionately affects indigenous women (De Chavez and Tauli-Corpuz, 2009; UNDP, 2009):

- Temperature increases between 2–4°C, especially in the Andean and the Caribbean regions, cause major changes in self-subsistence ecosystems. This affects traditional crop systems such as rotational agriculture, hunting practices and high mountain livestock. Women from Andean indigenous peoples such as Guambiano, Paez or Coconuco, can face food insecurity, energy shortage and an increase of health problems related to malnutrition. Negative impacts on traditional livelihoods and environments can cause a loss of traditional knowledge and cultural practices associated with indigenous peoples' cosmogonies and identity. Furthermore, peoples from the Amazon such as Uitoto, Ticuna or Andoque, and from the Pacific Coast such as Embera, are also highly impacted by the increase of temperature in rainforest ecosystems where flora and fauna are modified or destroyed, and malaria and dengue fever will increase.
- Reduction of precipitation by 30 per cent (mainly in the Andean region) decreases natural water springs and wells upon which indigenous women depend for their productive and reproductive activities. Reduction in rainfall has also resulted in the construction of dams and other megaprojects on high mountain rivers in order to guarantee the supply of clean water to populated cities. Such is the case in Sierra Nevada de Santa Marta (in the northeast of Colombia) where a water megaproject – the Besotes project – will affect sacred indigenous areas and also modify the natural irrigation system that supplies water to indigenous populations. Women from the Arhuaco, Kogi and Kankuamo peoples will face these negative consequences.
- Prolonged droughts and their impact on strategic high altitude ecosystems such as snow picks and uplands (75 per cent of upland plateaus and 95 per cent of snow picks are under risk of extinction) leads to the loss and destruction of ancestral environments. Rivers, forests (*bosques de niebla* – fog forests) and lagoons located in high

altitudes are essential for the ceremonial and ritual life of diverse peoples. Changes in these systems will also cause the destruction of crops, plants and species upon which indigenous communities base their livelihood. The Pastos women from Nariño (in the southwest of Colombia), who plant different kinds of potatoes at altitudes of more than 2950m, would experience a negative impact. Different species of highland potatoes are their main food source and their principal exchange product in the food market.

- More frequent and more severe natural disasters such as floods, landslides and desertification increase the vulnerability of indigenous women. Their physical isolation, their lack of resources to mitigate negative impacts such as the destruction of houses and infrastructure (bridges, roads, etc), reduced opportunities to exchange and market food products, lack of information on disaster preparedness and mitigation, and lack of health services put them and their children at high risk that is further increased by a natural disaster.

- Together with armed conflict, climate change is the main potential cause for the displacement of indigenous women. Indigenous women who are environmental refugees would likely be moving into urban contexts that are not prepared to host them. They would face lack of housing, water, sanitation, energy and waste management services. The UN Special Rapporteur on the Situation of Human Rights and Fundamental Freedoms of Indigenous Peoples (Anaya, 2009) has confirmed information and figures provided by Stavenhagen (2004) who stated that 60 per cent of displaced indigenous women and their children living in cities did not meet their health needs and suffered conditions such as diarrhoea, respiratory problems and malnutrition. This is particularly the case of indigenous women living in cities such as Bogotá, Valledupar, Cali or Cartagena.

Adaptation and mitigation strategies developed by indigenous women

Indigenous women put into practice traditional specialized knowledge to cope with climate change impacts. Intercropping and multiple cropping practices, planting crop varieties more resistant to drought or floods, finding alternative irrigation systems and taking special care of springs, wells and rivers through reforestation, are the most common strategies developed by women.

By emphasizing climate change and environmental sustainability as the central issues within their local organizations and national and international networks, indigenous women have also called the attention of policy-makers and authorities to these issues. Women's active political participation is crucial to prevent actions such as monocrops or deforestation that would worsen negative environmental impacts of climate change. Arhuaco and Kankuamo women are currently leading campaigns to include environmental and climate change issues in national policies and plans directed at indigenous peoples. It is the case of the recent conflict mitigation/ prevention plans ordered by the Constitutional Court (Auto092 and Auto004[1]) in 2009.

Furthermore, at the international level, indigenous women have actively participated in the creation of a Latin American Indigenous Forum on Climate Change. This was proposed by indigenous women in March 2009 at the Latin American Summit on Climate Change and Impacts on Indigenous Peoples (2009, p3) that took place in Lima. Addressing gender impacts, making visible women's possible contributions to sustainability and enforcing national and legal platforms that ensure their rights to a healthy territory are the main goals to be achieved by this forum.

Lessons learned and recommendations

Indigenous women have historically preserved biodiversity and have balanced ecosystems. They are exemplary environmental caretakers and users of non-contaminating productive techniques and sustainable methods to manage natural resources. Through traditional knowledge systems they guarantee cultural diversity, the cultural integrity of their peoples, and the survival of many flora and fauna species that are at risk of extinction. Learning from these cultural practices and recognizing their potential contribution is crucial to mitigate climate change's negative impacts on indigenous peoples. Documenting traditional practices and at the same time safeguarding their intellectual property rights, promoting workshops at the local, regional and global levels with indigenous women, and including interculturality as an important perspective to be mainstreamed into plans and policies of climate change are ways to achieve this goal.

Multilateral agencies, regional organizations and national and municipal governments should ensure the active participation of gender-balanced indigenous peoples' representations in their climate change policy planning processes. By taking into account the UN Declaration of Rights of Indigenous Peoples, national governments such as Colombia's can ensure better and more effective implementation of climate change mitigation measures. Likewise, adaptation funds should be designated and accessible for indigenous peoples, especially for indigenous women who are facing negative environmental impacts of climate change.

Notes

1   The term 'Auto' here refers to legal acts and mandates ordered by the Colombian Constitutional Court to the Colombian national government. See also www.corteconstitucional.gov.co.

References

Anaya, J. (2009) 'Informe preliminar del relator especial de Naciones Unidas sobre Pueblos Indígenas' www.indepaz.org.co/index.php?view =article&id=304%3Acomunicadoinforme-preliminar-relator-especial-james-anaya&option=com_content&Itemid=91, accessed November 2009

CODHES (Consultaría para los Derech Humanos y el Desplazmiento) (2004) *Boletín de la Consultaría para los Derechos Humanos y el Desplazamiento: Las Mujeres en la Guerra: De la Desigualdad a la Autonomía Política*, April, no 48, Bogotá, Colombia

DANE (2009) *Resultados Censo Nacional de Colombia*, www.dane.gov.co/ daneweb_V09, accessed 12 January 2010

De Chavez, R. and Tauli-Corpuz, V. (eds) (2009) *Guide on Climate Change and Indigenous Peoples*, 2nd edition, Tebbteba Foundation, Philippines

ECLAC (2006) *Género, Pobreza, Raza y Etnia: Estado de la Situación en América Latina*, ECLAC, Santiago de Chile

Latin American Summit on Climate Change and Impacts on Indigenous Peoples (2009) 'Declaración por la vida de la madre naturaleza y humana', Lima, 24–25 March 2009, unpublished document

Oxfam (2009) *Violencia Sexual en Colombia: Un Arma de Guerra*, Oxfam, Bogotá, Colombia

Stavenhagen, R. (2004) *Informe del Relator Especial de Naciones Unidas para los Pueblos Indígenas: Situación de Derechos Humanos y Libertades Fundamentales de los Indígenas: Misión Colombia*, Naciones Unidas, Bogotá, Columbia

UNDP (United Nations Development Programme) (2009) *Ficha de Cambio Climático para Colombia*, www.cambioclimatico.gov.co/

documentos/DocRefCambioClimatico/DocsEspanol/Memorias%20
Curso%20CC%20PNUD/Ficha%20cambio%20clim%C3%A1tico%20
Colombia.%20PNUD.pdf, accessed 4 December 2009

UN Permanent Indigenous Forum (2004) www.un.org/esa/socdev/unpfii/, accessed 1 December 2006

World Bank (2004) 'The gap matters: Poverty and well-being of indigenous peoples and Afro-Colombians, http://publications.worldbank.org/ecommerce/, accessed 24 November 2004

CASE STUDY 5.7

GENDER ASPECTS OF CLIMATE CHANGE IN THE US GULF COAST REGION

Rachel Harris

Introduction

The US Gulf Coast – that extends from Florida, via Georgia, Mississippi and Louisiana towards the eastern coast of Texas – became a worldwide media scene after the events of 29 August 2005. On that date, Hurricane Katrina made landfall there, the aftermath of which still affects the area five years later. Though much publicity focused on New Orleans, Louisiana, several other communities were impacted including in Gulfport and Biloxi in Mississippi and Houma in Louisiana. Hurricane Katrina was soon followed by Hurricane Rita, a more intense storm that hit areas west of those affected by Hurricane Katrina. What these storms demonstrated to the world is that the US is vulnerable to what some would call climate-related disasters. Global warming is continuing to raise sea surface temperatures which, in turn, expand the intensity, lifespan and frequency of hurricanes in the Atlantic. In fact, the year of 2005 broke records for the number of hurricanes in one season as well as the most intense and enormous hurricanes ever on record. In the aftermath of Hurricanes Katrina and Rita, there was much research done to assess the situation and the effects on the communities in the Gulf Coast. One reoccurring factor in post-storm research and reports was that women, especially poor African-American women, were hit the hardest and had the toughest time recovering from the disaster. Most of the research was done in the hardest hit areas of the Gulf Coast, including southern Louisiana and southern Mississippi. These are the areas on which this case study will focus.

Physical vulnerability

Physically, the coastlines of Louisiana and Mississippi are extremely fragile and vulnerable. In the past, the Louisiana coastline was covered with wetlands, natural barrier reefs and natural levees that were once barriers against flooding and high water surges. These have been depleted and eroded over the past several years. Erosion and destruction of wetlands are due to natural and human processes. Human-induced processes, including dredging for canals and unnatural levee systems along the Mississippi River, have increased erosion and decreased sediments that help to build up wetlands in southern Louisiana. It is estimated that Louisiana loses an area of wetlands the size of a football field every 38 minutes (Lindquist et al, 2007). When a storm comes into the Gulf Coast these areas become even more depleted. It is estimated that Hurricanes Katrina and Rita transformed 217 square miles of marsh to open water in coastal Louisiana (USGS, 2006). More recently, a study concluded that oil and gas pipelines are also contributing to depleting critical wetland areas (Johnston et al, 2009). The United Houma Nation, the largest indigenous tribe in Louisiana, which consists largely of a fishing community, has noticed saltwater intrusion because of the location of the canals used by the oil and gas industry to distribute oil to different places (Dardar-Robichaux, 2009).

Women's vulnerability

The combined factors of pervasive gender inequality and race discrimination, high poverty rates, low-wage jobs and large numbers of female-headed households in the Gulf Coast contributed and still contribute to women's vulnerability to disasters in the region.

In general, women in the southern region of the US are more likely to be poorer and lack health insurance and are less likely to be gainfully employed when compared to women in other parts of the US (Gault et al, 2005). Women in this area are largely African-American and experience discrimination based on both race and sex. The hardest hit states by the hurricanes were Louisiana and Mississippi; both rank bottom in the nation among indicators of women's status according to the Institute for Women's Policy Research.

Particularly female-headed households faced very high poverty rates in the Gulf Coast regions (Gault et al, 2005). In New Orleans and the metropolitan area of Gulfport/Biloxi, Mississippi, two of the areas most impacted by Hurricane Katrina, women faced much higher poverty rates than the national average with 25.9 per cent and 18.6 per cent respectively living below the poverty line, compared to 14.5 per cent nationally.

Many of the poor (about 44 per cent) are working at low-wage jobs. The percentage of the working poor is high in the Gulf Coast region, indicating that many employers do not provide enough hours or dollars to allow persons a way out of poverty. The high poverty rates of female-headed households may partially be due to a single parent working at a low-wage job.

Race and sex discrimination are still pervasive in the workplace and such disparities are obvious in the Gulf Coast region, where women and persons of colour are more likely than white men to work low-wage jobs. In New Orleans in particular, men out-earn women in nearly every occupation. When comparing women across races, it was found that when white women and black women had the same occupation, black women earned less than white women, sometimes substantially so (Williams et al, 2006).

In several Gulf Coast areas, the rebuilding process involved a lot of new construction, an occupation dominated by men. No child care was available, limiting the mobility of female-headed households. Public housing developments were torn down, even if they were not damaged, leaving very little affordable housing. According to the Greater New Orleans Community Data Center, renters and homeowners are currently facing unaffordable housing costs as a result of the storms of 2005. Interviews of women from the Gulf Coast found that the most pressing needs reported to begin recovery for women were jobs and housing: people who went back to work had a job and a purpose to get through the aftermath of the storm (Davis and Land, 2007).

Women activists after the storm: Anecdotes

Despite their poverty, the race divide and the environmental disasters, women in the Gulf Coast have been leaders in rebuilding and empowering other women to be more involved in the rebuilding process in the aftermath of the storms. After interviewing some of these leaders – who represent significant communities of women – they seemed to agree that the storm has had three major positive impacts on their communities:

- Greater involvement of women, especially low-income women, in community activism, rebuilding and recovery efforts.

- More organizations that did not work together in the past are now coming together for a common cause; many of these organizations are being led or started by women who have stepped up for the cause.
- More women are elected officials and are key stakeholders in policy and decision-making. However, although more women are in power they are not necessarily gender sensitive and the public face of politics is still very male-dominated.

According to Sara K. Gould and Cynthia Schmae of *Women's eNews*, women's funds were quickly expedited to the affected areas after Hurricane Katrina. Much of the initial funding was given to various grassroots organizations (Gould and Schmae, 2007). The Ms Foundation started the Katrina Women's Response Fund intended to support low-income women and women of colour in the Gulf Coast to ensure prioritization of their needs during rebuilding and recovery. Funds in many of the states where evacuees went also helped women evacuees especially, in readjusting their lives and their families to a new place in a time of crisis.

Such targeted funding seems to have helped women have a greater role in rebuilding and recovery and put informal and formal community networks into place in the aftermath of the disasters. Women such as Sharon Henshaw of Biloxi, Mississippi, saw the need for community activism in the aftermath of the storms (Henshaw, 2009). She brought a community of women together for weekly meetings, which eventually grew into the non-profit organization she leads today, Coastal Women for Change (CWC). She states: 'Men migrated away to get a job somewhere else following the disaster… [Biloxi has a population of] more women than men since the storm… [therefore] women are pushed to become empowered.' However, they do not stay empowered in the presence of men. CWC works both on national and local legislation to protect the communities from future disasters. Henshaw believes that the new Gulf Coast City Works Act, which has an agenda to create green jobs in Biloxi will allow women to take advantage of a green jobs opportunity 'because women are at the table'. Women have been the key to recovery in Biloxi and, therefore, have their voices heard in support of their needs and the needs of women and men in Biloxi.

These sentiments were also expressed by Ann Yoachim, Programme Manager of Tulane University's Institute on Water Resource

Law and Policy in New Orleans (Yoachim, 2009). Able to come back to the city within the first few months after the hurricane hit, she saw a city of mainly men (construction, contractors and National Guard presence), very few women and no children. She says the city was 'very male dominated and there were no children right after the storm... [this] caused it to not feel like a city, home or community'. It was when women and children began returning to the city that community-led recovery and collaboration took root.

Despite larger participation of women in community organizing following the storms, Shana Griffin, founder of the Women's Health & Justice Initiative (WHJI), points out that policy-makers and organizers have not utilized the recommendations of gender and disaster studies (Griffin, 2009). She has not witnessed a change in attitude regarding the importance of centring women in the decision-making process nor has she seen people examining gender issues during the recovery process. However, more organizations are catering to their constituencies to better prepare them for future disasters. At WHJI, they have created sexual and reproductive health disaster preparedness kits and increased research and awareness on the rise of HIV/AIDS cases occurring after disasters. However, as far as progress with policy, while she sees more organizations advocating for more wholesale solutions and policies and more women stakeholders, she does not witness much advocacy on behalf of the needs of women. This is something she would like to see change.

Brenda Dardar-Robichaux is, Principal Chief of the United Houma Nation (UHN), a matriarchal Native-American tribe in southwest Louisiana. The majority of persons within the UHN are women, and in times of hurricane recovery and/or changes in the community women are primarily responsible for making families feel safe again and rebuilding homes (Dardar-Robichaux, 2009). Traditionally in the UHN, men have been the fishers in the communities and are often out earning a living to provide for the family, leaving the women in the role of home recovery. However, saltwater intrusion has made fishing harder for the men and a majority of the younger generation of the UHN works on offshore oil rigs. Offshore work is hard and men often come home with health problems after being offshore for 7–14 days out of the month. Women are then left with a greater burden of caring for the family.

Following the hurricanes there has been a necessity for UHN women to go into the workforce, changing the traditional roles of the family. This has been a difficult transition as most women have not previously worked outside of the home and jobs are scarce. Climate changes impact greatly on the roles of women and men in the tribal communities. As a woman leader, Dardar-Robichaux has worked to develop a three-step plan to try to ease the burden on families in the UHN. This plan includes evacuation, making homes resilient for persons who stay and easing the process of relocation for persons who leave.

One of the major problems for the UHN following Hurricanes Katrina and Rita resulted from the fact that the UHN is not recognized as a legitimate tribe by the federal government, only by the state. Therefore, in the immediate aftermath of the hurricanes, the UHN received very little help nationally and had to recover with little resources. In light of this, the UHN is developing a hazard mitigation plan that would allow it to apply for federal assistance programmes, which provide funds for disaster preparedness and long-term recovery and rebuilding.

Conclusions

The gender differences in the Gulf Coast became practically a determinant of who could move back to the disaster-ridden areas, who could get jobs and how well persons could recover. Women were often left poorer, less able to find jobs and, therefore, less able to move back to their previous homes following the storm. However, as many have mentioned, this reality has also caused women who are in the recovery areas to become more involved and more active in recovery and community efforts. There has been a greater community of people working together and women are stakeholders in many of the decisions being made around recovery and rebuilding. That does not mean gender considerations are well taken into account. In fact, though there are more women leaders, decisions are still being made that do not always consider the different roles and burdens that women and men have to bear. It is the hope that future policies in the Gulf Coast and national policies begin to heed recommendations of the gender and disaster vulnerability research that have come out in the aftermath of Hurricanes Katrina and Rita. It is the only way to prevent another disaster of this magnitude and make adaptation and disaster preparedness efforts effective in the short and long-term.

The author wants to thank Brenda Dardar-Robichaux, Shana Griffin, Sharon Henshaw and Ann Yoachim for their cooperation.

References

Dadar-Robichaux, B. (2009), interview, 11 November 2009

Davis, O. A. and Land, M. (2007) 'Southern women survivors speak about Hurricane Katrina, the children and what needs to happen next', *Race, Gender & Class*, vol 14, nos 1–2, pp69–86

Gault, B., Hartmann, H., Jones-DeWeever, A. Werschkul, M. and Williams, E. (2005) *The Women of New Orleans and the Gulf Coast: Multiple Disadvantages and Key Assets for Recovery Part I: Poverty, Race, Gender and Class*, Institute for Women's Policy Research, D464, October 2005

Gould, S. K. and Schmae, C. (2007) 'Katrina's gender focus offers lessons for recovery', *Women's* eNews, August 2007, www.womensenews.org/print/6474, accessed 8 February 2010

Griffin, S. (2009) interview, 13 November 2009

Henshaw, S. (2009) interview, 12 November 2009

Johnston, J. B., Cahoon, D. R. and La Peyre, M. (2009) *Outer continental shelf (OCS)-related pipelines and navigation canals in the Western and Central Gulf of Mexico: Relative impacts on wetland habitats and effectiveness of mitigation*, US Department of the Interior, Minerals Management Service, Gulf of Mexico OCS Region, New Orleans, OCS Study MMS 2009-048

Lindquist, D. C. and Martin, S. R. (2007) *Coastal Restoration Annual Project Reviews: December 2007*, Louisiana Department of Natural Resources, Baton Rouge

USGS (US Geological Survey) (2006) *USGS Reports Latest Land Change Estimates for Louisiana Coast*, US Department of the Interior, www.usgs.gov/newsroom/article.asp?ID=1568, accessed 8 February 2010

Williams, E., Soronkina, O., Jones-DeWeever, A. and Hartmann, H. (2006) *The Women of New Orleans and the Gulf Coast: Multiple Disadvantages and Key Assets for Recovery Part II: Gender, Race, and Class in the Labor Market*, Institute for Women's Policy Research, D465, August 2006

Yoachim, A. (2009) interview, 13 November 2009

CASE STUDY 5.8

WOMEN AT WORK: MITIGATION OPPORTUNITIES AT THE INTERSECTION OF REPRODUCTIVE JUSTICE AND CLIMATE JUSTICE – EXAMPLES FROM TWO INDUSTRIAL SECTORS IN THE US[1]

Ann Rojas-Cheatham, Dana Ginn Paredes, Aparna Shah, Shana Griffin and Eveline Shen

Introduction

Effectively solving the climate crisis demands that the mitigation and adaptation measures employed align with a justice agenda that improves the circumstances of poor people, people of colour, women, and children. If synergistic efforts to protect the planet and to improve the lives of the most vulnerable among us are made, we will create a sustainable system that asks more of those with the most to give and less of those with least to spare. There is no doubt that, in order to solve the climate crisis, a new economic and political system that is both sustainable and just will need to be constructed.

Women – who have and will continue to bear an increasingly disproportionate share of the climate change burden in coming decades – are central to the success of constructing this new system. The current working paradigm regarding women and climate change focuses on the fact that women, specifically women of colour, are disproportionately impacted by disasters and environmental degradation caused by climate change. In the US, women are 45 per cent more likely to be poor than men (National Women's Law Center, 2006). Low-income women, immigrant women and women of colour will be most impacted by the severe weather events, heatwaves and increases in disease rates that will characterize climatic changes. This paradigm encourages interventions to increase women's capacity to adapt to these changes. While a focus on women and adaptation is a vital undertaking, steps must also be taken to expand our comprehension of how women are affected by climate change, its causes and by the solutions to mitigate it. In this case study, the current working paradigm regarding gender and climate change will be expanded by a framework that addresses an intersection of reproductive justice and climate justice. This new framework brings an understanding of the need to include strategies that improve the health and working conditions of women of colour working in low-wage toxic industries as part of climate change adaptation and mitigation. This approach allows us to more clearly identify mitigating solutions that advance both reproductive justice and climate justice.

The grassroots community-based organization Asian Communities for Reproductive Justice (ACRJ) has begun to develop and work within the intersection of reproductive justice and climate justice (ACRJ and WHJI, 2009). Before working within this intersection, ACRJ developed a reproductive justice framework that is described in detail in *A New Vision for Advancing Our Movement for Reproductive Health, Reproductive Rights and Reproductive Justice* (ACRJ, 2005).[2] Reproductive justice in the workplace includes having a healthy and safe work environment, access to health care, freedom from discrimination and the ability to earn a living wage with dignity and respect. Climate justice addresses the inequalities caused by climate change. Climate justice ensures the freedom and equality of all people by addressing the unequal oppression created and/or exacerbated by climate change, such as sexism, racism, classism and xenophobia.[3]

Healthy workplaces: Healthy women, healthy Earth

When looking at the causes that both contribute to climate change and harm the reproductive health of women of colour workers in low-wage toxic industries, an opportunity transpires to identify and implement strategies that advance both reproductive justice and climate justice. Because women of colour tend to work in mid-market industries (Ortiz, 2006) a focus on such industries is called for in order to identify strategic opportunities for working at that intersection.

The importance of mid-market industries

Efforts to mitigate climate change have focused on energy producing industries and the transportation sector. It is also critical to investigate secondary industries that depend on fossil fuel energy production.

These industries that have indirect or 'secondary' greenhouse gas (GHG) emissions, and collectively are as dirty as the top emitters, are called mid-market companies (David and Lucile Packard Foundation et al, 2007). Reducing the emissions of mid-market companies has been identified as one of the top five most important strategies to reduce global warming by multiple experts (David and Lucile Packard Foundation et al, 2007). If mid-market industries don't alter their products and demands for energy, oil refineries and coal plants that are major sources of CO_2 emissions will continue to produce the same supply.

Understanding the life cycle of chemicals

Though mid-market industries do not directly emit significant amounts of GHGs, they have a greater than expected impact on global warming through GHG emissions released in the full life cycle (extraction, production, distribution, use and disposal) of the primary chemicals and materials used to make their products. Life cycle assessments (LCAs) can provide measurements of the impact of the entire life cycle of a chemical or material on particular environmental aspects such as energy consumption, GHG emissions or water contamination. LCA is a way to measure the GHG emissions from the products and supplies used by industries on top of the emissions from the production process.[4]

In the following low-wage industries with predominantly female workforces, climate change mitigation opportunities can also improve reproductive justice: the semiconductor industry and the nail salon industry.

The semiconductor industry

The semiconductor assembly industry[5] is a low-wage top-emitting industry that primarily employs women of colour. The industry is classified by the US Environmental Protection Agency (EPA) as one of the top six industrial processes that contribute to global warming (US EPA, 2010). Current semiconductor manufacturing processes require the use of high global warming potential (GWP) fluorinated compounds, including perfluorocarbons, trifluoromethane, nitrogen trifluoride and sulfur hexafluoride, collectively termed perfluoro compounds (PFCs) (US EPA, 2009). PFCs have been identified as some of the most potent GHGs measured (US EPA, 2010).

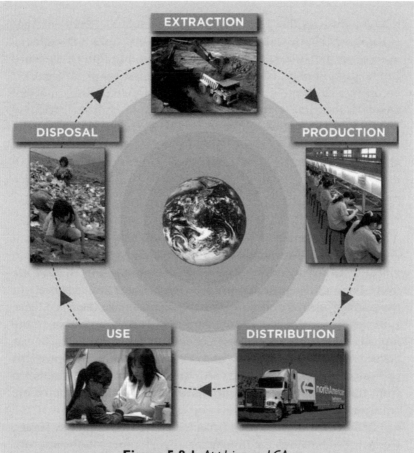

Figure 5.8.1 *Applying an LCA*

Source: ACRJ and WHJI, 2009, p13

While processes used in the semiconductor industry significantly contribute to global warming, the sector also considerably harms reproductive health and justice for women workers. Research has demonstrated that women working in the semiconductor industry may have an increased risk of delivering a low birth weight baby (Lipscomb et al, 1991), congenital malformation (Saillenfait and Robert, 2000), spontaneous abortion and subfertility (Correa et al, 1996; Schenker et al, 1995), cancer (Chen, 2007) and musculo-skeletal problems (Chee and Rampal, 2004). Initial research has found associations between these health effects and workers' exposure to chemicals, including PFCs, used to manufacture microchips

(Smith et al, 2006). Comprehensively defining reproductive and cancer risks of women working in the semiconductor industry has been difficult because the US semiconductor industry has not supported access for independent studies (LaDou and Bailar, 2008).

In the case of the semiconductor industry, if we would identify the reproductive health impact as the sole focus of workplace change, then it could be determined that a better ventilation system is needed to decrease workers' exposure to chemicals. But when we look both ways and in addition to reproductive justice take into account the semiconductor industry's contribution to global warming, we are prompted to seek more comprehensive solutions that eliminate or reduce the hazardous chemicals that are harmful to women workers and the environment with specific attention towards decreasing GHG emissions.

The nail salon industry

The second industry this case study will apply the reproductive justice and climate justice framework to is the nail salon industry. Nail care is the fastest growing sector in the beauty industry, generating more than US$2.8 billion in sales annually in the US in 2000 (US Census Bureau, 2002).[6] These salons provide a critical source of employment for women of colour. According to Federmann et al (2006) in California alone, there are approximately 8300 nail salons and more than 300,000 people licensed to work in them. The majority of the nail salons in California are owned and staffed by Vietnamese women (Federmann et al, 2006).

Nail salons use a large range of unregulated (in the US) chemicals in products for nails as well as products mandated for use in cleaning the salons. The chemicals used include solvents, hardeners, fragrances, glues, polishes, and dry/curing agents. In addition to the chemicals in the cleaning products they are required to use, these chemicals pose risks to the health of nail salon workers, the environment and the climate. Climate change contributors used in nail salons are acetone, aluminium, ammonia, petrochemicals, magnesium and phenols. If life cycle assessments of the chemicals used in nail products and cleaning materials were to be conducted, the assessment of the industry's GHG emissions would substantially increase as compared to figures in the traditional use analysis.

Volatile organic compounds (VOCs) are chemicals used in nail salon products that contribute to global warming through their role in the formation of smog. Acetone, a solvent in nail polish remover, is an example of one of the VOCs emitted directly from the use of remover in the salon. Many nail polishes contain the chemical aluminium calcium sodium silicate which requires aluminium to produce. The production of aluminium is one of six industrial processes classified as having high GWP by the EPA (US EPA, 2010) because during primary aluminium production, PFCs are emitted as byproducts of the smelting process. Therefore, the production of nail polish is dependent upon one of the top six GWP gas emitting industrial processes.

The production of ammonia, used to clean nail salons, is the 14th largest source of CO_2 in the US (US EPA, 2009b). Petrochemicals are used in nail polish in the form of polyvinyl chloride; their production is the 18th largest source of CO_2 in the US (US EPA, 2009b). Magnesium is also found in nail polish, and magnesium production and processing is the second largest source of sulfur hexafluoride in the US. Sulfur hexafluoride has one of the largest GWP of all GHGs (US EPA, 2009b).

In the US, nail salons are required to use hospital-grade cleaners and disinfectants. Chemicals known as phenols are used in nail salons as disinfectants, degreasers and slimicides to clean the equipment and the salons themselves. In order to produce phenols, large amounts of the GHG nitrous oxide (N_2O) are used. Manufactured sources of nitrous oxide accounted for approximately 4.4 per cent of all GHG emissions in the US in 2007 (US EPA, 2009c). Although N_2O emissions are much lower than CO_2 emissions, N_2O is approximately 300 times more powerful than CO_2 at trapping heat in the atmosphere and nitrous oxide production is the third largest source of N_2O emissions in the US (US EPA, 2009c).

The impact of the use and disposal of nail salon products also creates GHG emissions. For example, at an ex-Revlon site in New Jersey that was used to manufacture nail salon products, tetrachloroethene (PCE) and other VOCs have been found to be serious polluting contaminants (Amuthan, 2008). The health effects of PCE in humans include neurological, liver and kidney problems following short-term and long-term inhalation. PCE evaporates readily from soil and surface water and undergoes degradation in air to produce direct and indirect GHGs that contribute to climate change including phosgene, trichloroacetyl chloride, hydrogen chloride, carbon monoxide and CO_2.

Reproductive justice is only possible when women are physically healthy and when their economic, political and social rights are assured. Although more longitudinal research is needed in order to fully understand the long-term health impacts of working in a nail salon, the research to date points to possible serious health effects. Studies have shown that manicurists and cosmetologists may experience disproportionate rates of multiple myeloma (Guidotti et al, 2007), spontaneous abortion (John et al, 1994), birth defects, reproductive problems and asthma (Porter, 2009). Moreover, a reproductive justice analysis of working conditions in nail salons directs improvements not only to making the nail salon environment one that is conducive to good health, but also to increasing wages, improving benefits, reducing working hours, reducing harassment and discrimination, and creating more educational opportunities for the workers.

POLISH: Nail salon workers looking both ways

ACRJ has applied the reproductive justice and climate justice framework to the nail salon industry, focusing specifically on nail salon workers in Oakland, California through the project POLISH (Participatory Research, Organizing and Leadership Initiative for Safety and Health) that organizes workers. POLISH is a leading member of the California Healthy Nail Salon Collaborative, an alliance of community, advocacy, policy and research organizations dedicated to advancing a preventative environmental health agenda for the nail salon industry in California. Over the past two years, this project has made gains in a successful campaign pressuring government agencies to prioritize education and access to health and safety information for workers and salon owners. Currently, POLISH is designing a local campaign in Oakland to improve reproductive justice for nail salon workers and reduce the GHGs emitted by products used in this industry.

Conclusions and lessons learned

In the US, both the semiconductor and nail salon industries employ significant numbers of immigrant women; both industries have been found to have possible negative reproductive health consequences for workers and both are, in their own way, complicit in the widespread chemical production and consumption that contribute to climate

Figure 5.8.2 *Industry change that can mitigate climate change and improve the health of workers*

Source: ACRJ and WHJI, 2009, p18

change. The nail salon industry represents a mid-market industry that experts agree needs to be comprehensively addressing the climate crisis. The semiconductor industry has been identified as a top GHG emitter. Solutions that combine reproductive justice and climate justice within these two industries represent new, local and forward-looking strategies to the climate crisis – ensuring both sustainability and justice. Similar solutions should be strived for in other low-wage toxic industries with primarily female workforces. The opportunity is ripe for corporations and individual businesses to be accountable to the reproductive health of the women whom their success depends on and at the same time reduce or eliminate GHG emissions.

These changes can happen at multiple levels. Entire industries as well as individual businesses can take action at the intersection. Individual businesses can choose to change and/or eliminate hazardous products. Cities and regions can also provide incentives and support local businesses to make changes that advance reproductive justice and climate justice.

Change at the industry level can happen as a result of government regulations and/or voluntarily. For example, an industry can take pre-emptive actions or be regulated to change the chemicals and materials in the production that impact on reproductive health and global warming. Cities, counties and states can participate in creating an environment that supports industries to transition by rewarding those that incorporate greener and safer practices. In order for cities and/or states to make the link from products and chemicals used to global warming, several steps are important. First, they must incorporate an LCA framework to measure GHG emissions; thereby making tools and methods available to conduct LCA and analyse product and supply chains. Secondly, government and industry must collaborate with affected communities and incorporate the leadership and solutions from workers themselves. Ultimately through this range of government and community activity, it will be more probable for industry to be on a path towards altering the chemical composition of what they supply. This will result in healthier outcomes for workers, surrounding communities and the planet.

Moving beyond a gender analysis that solely focuses on the disproportionate impact of climate change on women, and embracing the complex interactions between women's well-being and climate change mitigation holds the potential to activate and mobilize larger constituencies that advance reproductive justice and climate justice and keep our movements strong, relevant and forward-looking.

Notes

1 This case study is based on a report by Asian Communities for Reproductive Justice (ACRJ) and the New Orleans Women's Health & Justice Initiative (WHJI) entitled *Looking Both Ways: Women's Lives at the Crossroads of Reproductive Justice and Climate Justice* by A. Rojas-Cheatham, D. Paredes, S. Griffin, A. Shah and E. Shen. Elements of the original report have been updated for purposes of this case study. http://reproductivejustice.org/assets/docs/ACRJ-MS5-Looking-Both-Ways.pdf, accessed 4 January 2010.

2 Reproductive justice exists when all people have the economic, social, and political power and resources to make healthy decisions about our gender, bodies and sexuality for ourselves, our families and our communities. This definition was developed by ACRJ (2005).

3 This definition of climate justice was developed by members of ACRJ's youth project, Sisters in Action for Issues of Reproductive Empowerment (SAFIRE). The members of SAFIRE are currently learning about climate change and are designing a project at the intersection of reproductive justice and climate justice. SAFIRE members first learned about climate justice as defined by the organization, Environmental Justice and Climate Change Initiative (EJCC) and then developed their own definition described in this case study.

4 It is very apparent to the authors of this case study that LCAs would significantly assess increased emissions estimates as compared to a traditional point of use or point of production analysis. This was confirmed in a conversation between Ann Rojas-Cheatham (primary author of this case study) and Daniel T. McGrath, Director of the Berkeley Institute of the Environment on 16 October 2009. In the University of California Berkeley Climate Change Feasibility Study, adding the life cycle calculation to the emissions inventory can be expressed as a 130 per cent emissions increase (UC Berkeley Climate Action Partnership Feasibility Study 2006–2007 Final Report, http://sustainability.berkeley.edu/calcap/docs/CalCAP%20Report%20FINAL%202007.pdf, accessed 10 January 2010).

5 The semiconductor industry is a multiple lever and technology enabler for the whole electronics value chain (www.en.wikipedia.org/wiki/Semiconductor_industry). While the semiconductor industry embodies the aggregate collection of companies engaged in the design and fabrication of semiconductor devices, the assertions of this paper are specific to the fabrication phase (manufacturing and production) of these products.

6 Nail salons generate US$2.8 billion per year according the US census data from the year 2002. This number was generated by adding receipts from employers (US$1000) figures and from non-employers (businesses with no paid employees) (US Census Bureau, 2002).

References

ACRJ (Asian Communities for Reproductive Justice) (2005) *A New Vision for Advancing our Movement for Reproductive Health, Reproductive Rights, and Reproductive Justice,* http://reproductivejustice.org/assets/docs/ACRJ-A-New-Vision.pdf, accessed 3 January 2010

ACRJ and WHJI (Women's Health & Justice Initiative) (2009) *Looking Both Ways: Women's Lives at the Crossroads of Reproductive Justice and Climate Justice,* http://reproductivejustice.org/assets/docs/ACRJ-MS5-Looking-Both-Ways.pdf, accessed 4 January 2010

Amuthan, L. A. (2008) 'Serious contaminants among pollution at ex-Revlon site in Edison', www.mycentraljersey.com/article/20080819/NEWS/808190385/DEP--%5C-Serious-contamints%5C--among-pollution-at-ex-Revlon-site-in-Edison, accessed 20 January 2009

Chee, H. L. and Rampal, K. G. (2004) 'Work-related musculoskeletal problems among women workers in the semiconductor industry in Peninsular Malaysia', *International Journal of Occupational and Environmental Health,* vol 10, no 1

Chen, H. (2007) 'Exposure and health risk of gallium indium, and arsenic from semiconductor manufacturing industry workers', *Bulletin of Environmental Contamination and Toxicology,* vol 78, no 2, pp123–127

Correa, A., Gray, R. H., Cohen, R., Rothman, N., Shah, F., Seacat, H. and Corn, M. (1996) 'Ethylene glycol ethers and risks of spontaneous abortion and subfertility', *American Journal of Epidemiology,* vol 143, no 7, pp707–717

David and Lucile Packard Foundation, the Doris Duke Charitable Foundation, Energy Foundation, Joyce Foundation, California Environmental Associates, Oak Foundation, and the William and Flora Hewlett Foundation, (2007) 'Design to win, philanthropy's role in the fight against global warming', p7, www.climateactionproject.com/docs/Design_to_Win_8_01_07.pdf, accessed 3 January 2010

Federmann, M.N., Harrington, D.E. and Krynski, K. (2006) 'Vietnamese manicurists: are immigrants displacing natives or finding new nails to polish?', *Industrial and Labor Relations Review,* no 59, pp302–318

Guidotti, S., Wright, W. and Peters, J. (2007) 'Multiple myeloma in cosmetologists', *American Journal of Industrial Medicine,* vol 3, no 2, pp169–171

John, E., Savitz, D. and Shy, G. (1994) 'Spontaneous abortions among cosmetologists' *Epidemiology,* vol 5, no 2, pp147–155

LaDou, J. and Bailar, J. (2008) 'Cancer and reproductive risks in the semiconductor industry', *International Journal of Occupational and Environmental Health*, vol 14, no 2, p152

Limpscomb, J. A., Fenster, L., Wrensch, M., Shusterman, D. and Swan, S. (1991) 'Pregnancy outcomes in women potentially exposed to occupational solvents and women working in the electronics industry', *Journal of Occupational and Environmental Medicine*, May 1991, vol 22, no 5, pp585–653

National Women's Law Center (2006) 'Losing ground: An overview of poverty, income and health insurance trends among women 2000–2005', www.nwlc.org/pdf/IncomePoverty3.pdf, accessed 4 December 2009

Ortiz, P. (2006) 'Industries that are tops for women of color', *DiversityInc*, www.diversityinc.com/content/1757/article/318/?Industries_That_Are_Tops_for_Women_of_Color, accessed 3 January 2010

Porter, C. A. (2009) *Overexposed and Underinformed: Dismantling Barriers to Health and Safety in California Nail Salons Report and Policy Agenda*, California Healthy Nail Salon Collaborative Report, http://saloncollaborative.files.wordpress.com/2009/03/collaborativepolicyrecreport.pdf, accessed 10 January 2010

Saillenfait, A. and Robert, E. (2000) 'Occupational exposure to organic solvents and pregnancy: Review of current epidemiologic knowledge', *Revue d'Epidémiologie et de Santé Publique*, vol 48, no 4, pp374–388

Schenker, M., Gold, E., Eskenazi, B., Hammond, S., Lasley, B., McCurdy, S., Samuels, S., Saiki, C. and Swan, S. (1995) 'Association of spontaneous abortion and other reproductive effects with work in the semiconductor industry', *American Journal of Industrial Medicine*, vol 28, no 6, pp639–659

Smith, T., Sonnenfeld, D. and Naguib Pellow, D. (2006) *Challenging the Chip: Labor Rights and Environmental Justice in the Global Electronics Industry*, Temple University Press, Philadelphia

US Census Bureau (2002) 'Industry Statistics Sampler NAICS 812113: Nail Salons', www.census.gov/econ/census02/data/industry/E812113.HTM, accessed January 8 2010

US EPA (US Environmental Protection Agency) (2009a) 'PFC reduction/climate partnership for the semiconductor industry', www.epa.gov/semiconductor-pfc/basic.html, accessed 3 October 2009

US EPA (2009b) 'Inventory of US greenhouse gases emissions and sinks: 1990–2008, Chapter 4: Industrial Processes', http://epa.gov/climatechange/emissions/downloads09/GHG2007-04-508.pdf, accessed 9 January 2010

US EPA (2009c) 'Inventory of US greenhouse gases emissions and sinks: 1990–2008: Executive Summary', p11, http://epa.gov/climatechange/emissions/usgginv_archive.html, accessed 9 January 2010

US EPA (2010) 'Sources and emissions', *High Global Warming Potential (GWP) Gases*, www.epa.gov/highgwp/sources.html, accessed 3 January 2010

Part III
Strategies and Action

6

Establishing the Linkages between Gender and Climate Change Adaptation and Mitigation

Lorena Aguilar

This chapter aims to identify the gender specific aspects of climate change adaptation and mitigation. It reconfirms the need to not only follow a vulnerability approach in addressing climatic changes, but also to delve deeper into its risk factors. It presents strategic steps to promote gender equality in climate change adaptation and mitigation policies, programmes and activities.

Introduction

Over the past two decades, climate change has increasingly become recognized as a serious threat to sustainable development, with current and projected impacts on areas such as the environment, agriculture, energy, human health, food security, the economy, natural resources and physical infrastructure.

Although climate change will affect all countries, its impacts will be differently distributed among regions, generations, classes, income groups, occupations and genders (IPCC, 2001). The poor (of which the majority are women), living primarily but by no means exclusively in developing countries, will be disproportionately affected (Drexhage, 2006). Climate change does not affect women and men in the same way and it has, and will have, a gender-differentiated impact. Therefore, all aspects related to climate change (such as mitigation, adaptation, policy development and decision-making) must include a gender perspective. Unfortunately most of the debate related to climate change has been gender blind.

However, women are not just helpless victims of climate change – women are powerful agents of change and their leadership is critical. Women can help or hinder strategies that deal with issues such as energy consumption, deforestation, burning of vegetation, population and economic growth, development of scientific research and technologies, and policy-making, among many others.

Therefore assessments of the impacts of climate change and related policies and programmes should be comprehensive and take into account:

- Causes of vulnerability or specific conditions that make women, especially poor women, vulnerable to climate change (see also Chapters 2 and 3).
- Gender-specific risk assessment and risk management.
- Gender aspects of strategies for enhancing adaptive capacity and mitigation actions.

Causes of vulnerability or specific conditions that make women, especially poor women, vulnerable to climate change

There is a need to avoid being simplistic and seeing women as victims simply because of their sex. Women are not vulnerable because they are 'naturally weaker': women and men face different vulnerabilities due to their gender conditions. For example, many women live in conditions of social exclusion. This is expressed in facts as simple as differentials in the capacity to run or swim, or constraints in women's mobility and behavioural restrictions that hinder their ability to relocate without their husband's, father's or brother's consent.

It has also been found that the vulnerability and capacity of a social group to adapt to or change highly depends on their assets. Next to their physical location, women's assets such as resources and land, knowledge, technology, power, decision-making, education, health care and food have been identified as determinant factors of vulnerability and adaptive capacity. As pointed out by Moser and Satterthwaite (2008), the more assets people have, the less vulnerable they are and the greater the erosion of people's assets the greater their insecurity. Unfortunately data from around the world indicates that women tend to have less or limited access to assets (physical, financial, human, social and natural capital).[1]

Women's assets largely determine how they will respond to the impacts of climate change. Therefore, actions should be taken to build up the asset base of women as a fundamental principle in adaptation strategies.

Box 6.1 *Gender inequality and resources*

Gender inequality exists in the access to valuable resources such as land, credit, agricultural inputs, technology and extension and training services that would enhance their capacity to adapt. The National Biodiversity Strategy and Action Plan (NBSAP) of Liberia mentions that women produce 60 per cent of food crops despite their lack of access to farmland, low level of technological training and knowledge, and lack of financial assistance (Liberia NBSAP, 2004). An analysis of credit schemes in five African countries conducted by the UN Food and Agriculture Organization (FAO) (2008) found that women received less than 10 per cent of the credit awarded to male smallholders. Fewer than 10 per cent of women farmers in India, Nepal and Thailand own land (FAO, 2008). In Kenya, although their statutory laws do not prevent them from owning land, women still face numerous difficulties in trying to do so (Kenya NBSAP, 2000).

Gender-specific risk assessment and risk management

Climate change will lead to increases in several risks. The threats to household well-being will stem from both direct risks (changes in climate variables) and indirect risks (such as increase in prevalence of pest and diseases). The impacts are likely to be felt disproportionately by particular individuals depending on characteristics such as sex, race (as, for example, in the aftermath of Hurricane Katrina) and age. Also, climate change will not only exacerbate existing risks but also reveal new risks that have been hidden (See Table 6.1).

In 2007, the study carried out by the London School of Economics that analysed disasters in 141 countries provided decisive evidence that gender differences in deaths from natural disasters are directly linked to women's economic and social rights. That is, when women's rights are not protected, more women than men die as a result of disasters. The study also found the opposite to be true: in societies where women and men enjoy equal rights, disasters kill the same number of women and men (Neumayer and Plümper, 2007). This means that the empowerment of women should be one of the priorities in adaptation and risk reduction strategies.

Climate variability has played an important role in initiating malaria epidemics in the East African highlands (Zhou et al, 2004) and accounts for 70 per cent of the variation of recent cholera series in Bangladesh (Rodó et al, 2002). This increase in outbreaks could have gender-differentiated impacts because women have less access to medical services than men

(Nelson et al, 2002) and women's workloads increase as they have to spend more time caring for the sick. Pregnant women particularly are more susceptible to malaria, as they appear to be physiologically more attractive to mosquitoes (WHO, 2009a).

Therefore, interventions related to risk reduction and social risk management should pay special attention to the need to enhance the capacity of women to manage climate change risks with a view to reduce their vulnerability and maintain or increase their opportunities for development. Some possible actions are:

- Improving access to skills, education and knowledge.
- Improving disaster preparedness and management.
- Supporting women to develop voice and political capital to demand access to risk management instruments.
- Developing policies to help households stabilize consumption (credit, access to markets and social security mechanisms).

Gender aspects of strategies for enhancing adaptive capacity and mitigation actions

Women and men have showed that they have different strategies and play different roles while coping with climate change. Women have always been leaders in community revitalization and natural resource management. They have proved to be powerful agents of change.

There are plenty of examples where women's participation has been critical to community survival. In Honduras, La Masica was the only community to register no deaths in the wake of Hurricane Mitch (1998) due to an early warning system operated by women in the community (Inter-American Development Bank, 1999).

Women from indigenous and local communities, based on their traditional knowledge, are also in possession of repertoires of what are called 'coping strategies', which they have used to manage climate variability and change. For example, they have developed cropping system adaptation strategies such as the use of a diversity of crops and varieties. In Rwanda, for instance, women are reported to produce more than 600 varieties of beans; and Peruvian Aguaruna women plant more than 60 varieties of manioc (FAO, 2001). These vast varieties, developed by women over the centuries, allows them to adapt their crops to different biophysical parameters including quality of soil, temperature, inclination, orientation, exposure and disease tolerance, among others.

Other coping strategies developed by women are related to the production of livestock. For example in southeast Mexico, women keep

Box 6.2 *Examples of women's roles in climate change mitigation*

Men and women often have different roles with regard to forest resource management. They play different parts in planting, protecting or caring for seedlings and small trees, as well as in planting and maintaining homestead woodlots and plantations on public lands. Men are more likely to be involved in extracting timber. Women typically gather non-timber forest products (NTFPs) for commercial purposes and to improve the living conditions within their households (as medicines, for example, or fodder for livestock). Since 2001, under the Maya Nut Program supported by The Equilibrium Fund, women in Guatemala, Nicaragua, El Salvador and Honduras have planted 400,000 Maya Nut trees (*Brosimum alicastrum*). The Equilibrium Fund is trying to participate in carbon trading with the US and Europe to show how specific projects could help improve women's lives, adapt to changes caused by climate change and reduce greenhouse gases (GHGs) (The Equilibrium Fund, 2007).

In the UK, the National Federation of Women's Institutes (NFWI) and Women's Environmental Network (WEN) supported the Women's Manifesto on Climate Change. A survey shows 80 per cent of women are very concerned about climate change; more women than men want more green products and carbon labelling of goods (85 per cent); lower prices for environmentally friendly products (85 per cent); and more government grants and incentives for energy efficiency and micro-generation to reduce carbon emissions (NFWI and WEN, 2007). (See also Chapter 9, Case study 9.2.)

According to studies conducted by the Organisation for Economic Co-operation and Development (OECD, 2008a, b), gender has a huge influence on sustainable consumption, partly due to the differing consumption patterns of men and women. In some OECD countries, women make more than 80 per cent of the consumer decisions and women are more likely to be sustainable consumers. In the US, Hope to Action is taking into account women's role in household consumption decision-making by organizing 'Eco Salons', which gather together women in their neighbourhoods and communities and provide practical advice on how to mitigate climate change through sustainable consumption (Hope to Action, 2010). Women's empowerment is linked to climate change solutions. In November 2006, Kenya's Green Belt Movement – founded by Nobel Peace Laureate Wangari Maathai – and the World Bank's Community Development Carbon Fund, signed an emissions reductions purchase agreement to reforest two mountain areas in Kenya. Women's groups would plant thousands of trees, an activity that would also provide poor rural women with a small income and some economic independence. Women's empowerment through this process would also capture 350,000 tons of CO_2, restore soil lost to erosion, and support regular rainfall essential to Kenya's farmers and hydroelectric power plants (Green Belt Movement, 2009; UNEP, 2010).

Table 6.1 *Indirect and direct risks associated with climate change and their potential effects on women*

Climate change effects	Potential risks	Examples of potential risks	Potential effect on women
Direct	Increase in ocean temperatures	In 2005, coral in the Caribbean suffered a bleaching event due to thermal stress. (Donner, et al, 2007).	The tourism industry is a particularly important sector for women; 46 per cent of the workforce are women (Anker, 1998).
	Increase in droughts and water shortages	Globally, the negative impacts of future climate change on freshwater systems are expected to outweigh the benefits (high confidence). By the 2050s, the area of land subject to increasing water stress due to climate change is projected to be more than double that with decreasing water stress. Areas in which run-off is projected to decline face a clear reduction in the value of the services provided by water resources. Increased annual run-off in some areas is projected to lead to increased total water supply. However, in many regions, this benefit is likely to be counterbalanced by the negative effects of increased precipitation variability and seasonal run-off shifts in water supply, water quality and flood risks (high confidence) (IPCC, 2008).	Erratic climate shifts can lead to conflicts over food, fuel and water. In the Darfur region of the Sudan, where drought and desertification have increased so has destruction of homes and villages. Women in particular have been the victims of intimidation, rape and abduction forcing thousands to become refugees (van Cotthem, 2009). Women and (usually) girl children fetch water in pots, buckets or ideally more modern narrow-necked containers, which are carried on the head or on the hips. A family of five needs approximately 100 litres of water each day to meet its minimum needs; the weight of that water is 100kg. In these circumstances, women and children may need to walk to the water source two or three times each day, with the first of these trips often taking place before dawn. During the dry season in rural India and Africa, 30 per cent or more of a woman's daily energy intake is spent in fetching water. Carrying heavy loads over long periods of time causes cumulative damage to the spine, the neck muscles and the lower back, thus leading to the early ageing of the vertebral column (World Bank et al, 2009). A study on drought management in Ninh Thuan, Vietnam showed that 64 per cent of respondents agreed that recurring disasters have differential impacts on women and men, and 74 per cent of respondents believed that women were more severely affected by drought than men due to differing needs for water. Women collect water from water sources that are farther away as each drought takes its toll. With fewer water sources nearby, women have to walk long distances to fetch drinking water (Oxfam, 2006).

Table 6.1 (continued)

Climate change effects	Potential risks	Examples of potential risks	Potential effect on women
Indirect	Increase in extreme weather events	Increase in intensity and quantity of cyclones, hurricanes, floods and heatwaves.	In countries where gender inequalities are deeper, during the disasters the ratio of female to male deaths was 4:1 (Neumayer and Plümper, 2007).
	Increase in epidemics	El Niño/Southern Oscillation accounts for 70 per cent of variation of recent cholera series in Bangladesh (Rodó et al, 2002) and climate variability played an important role initiating malaria epidemics in the East African highlands (Zhou et al, 2004).	Each year, some 50 million women living in malaria-endemic countries become pregnant; half of them live in tropical areas of Africa with high transmission rates of the parasite that causes malaria. An estimated 10,000 of these women and 200,000 of their infants die as a result of malaria infection during pregnancy; severe malarial anaemia is involved in more than half of these deaths (WHO, 2009b).
	Loss of species	The International Union for Conservation of Nature (IUCN) Red List provides information on the status of the world's birds (9856 species), amphibians (6222 species) and reef-building corals (799 species), among other groups. Preliminary analyses of life history and ecological traits of these groups suggest that up to 35 per cent of birds, 52 per cent of amphibians and 71 per cent of reef-building corals have traits that make them particularly susceptible to climate change (IUCN, 2009).	Women often rely on a range of plant, animal and crop varieties (agrobiodiversity) to accommodate climatic variability, but permanent temperature change will reduce agrobiodiversity and traditional wild food and medicine options (FAO, 2008).
	Decrease in crop production	If temperatures rise by more than 2°C, global food production potential is expected to contract severely and yields of major crops like maize may fall globally. The declines will be particularly acute in lower-latitude regions. In Africa, Asia and Latin America, for example, yields could decline by 20–40 per cent. In addition, severe weather occurrences such as droughts and floods are likely to intensify and cause greater crop and livestock losses (IPCC, 2007).	Women produce more than 50 per cent of the food worldwide. In Africa the percentage of women affected by these changes could range from 48 per cent in Burkina Faso to 73 per cent in the Congo (FAO, 2008).

> Climate change has critically changed the relevance of the traditional knowledge of indigenous groups. Natural signals that were used to trigger activities in the past are now less reliable. For example, as the weather becomes hotter in the tropics, migratory birds come at a different time of the year and the rainy season comes earlier or later than usual, which can lead to a disorientation of people in their daily lives (Macchi et al, 2008)

as many as nine breeds of local hens, as well as breeds of ducks, turkeys and broilers in their back gardens selecting the best breeds for the local environmental conditions (FAO, 2006).

However, it is important to point out that even though women have developed important strategies to cope with or to adapt to changes, the magnitude of future hazards may exacerbate their capacity to adapt.

In mitigation actions, women are key actors in reducing emissions through adopting clean household energy sources and technologies. Fuel combustion, improved stoves, biogas digesters and solar cookers are examples of many new cleaner fuels and more efficient energy solutions that can reduce emissions and provide other benefits to women: reducing indoor air pollution and freeing time they spent collecting and transporting biomass fuels to spend on new activities, such as education, production and economic advancement.

In forest resource management, in climate change mitigation measures aimed at increasing the storage of GHG, men and women often have different roles. They play different parts in planting, protecting or caring for seedlings and small trees, as well as in planting and maintaining homestead woodlots and plantations on public lands. Men are more likely to be involved in extracting timber and NTFPs for commercial purposes. Women typically gather forest products for fuel, fencing, food for the family, fodder for livestock and raw materials to produce natural medicines, all of which help to increase family income (Aguilar et al, 2007).

Men and women have different bodies of knowledge, skills and experience related to forest resources and they drive different benefits from forestry services. Significant disparities exist in access of women and men to forest-related decision-making, institutions and economic opportunities. In many countries, women do not have the same legal ownership rights, access to or control of forests, even though their livelihoods depend on forest resources. In fact women lack *de jure* right over land in many developing countries; however, they play a significant role in forestry as *de facto* users.

This gender differentiation of roles, responsibilities, rights, priorities and needs, is especially highlighted by climate change financial mechanisms directed at forestry. An analysis of several community forest groups in India and Nepal highlighted the fact that, in most cases, cash is not distributed equally and funds are commonly invested in resources or activities from which women were unlikely to benefit, such as club repair, purchasing community utensils, rugs, drums, etc. (Agarwal, 2002).

Financial mechanisms

It is important to understand the opportunities and dangers for gender equality presented by the new financial mechanisms. Reducing Emissions from Deforestation and Forest Degradation (REDD) and REDD+ offer incentives for developing countries to reduce emissions from forested lands and invest in low-carbon paths to sustainable development.[2] It is predicted that financial flows for GHG emission reductions from REDD could reach up to US$30 billion a year (REDD, 2009). Such amounts of funding can significantly contribute to the empowerment of women or, on the contrary, can deepen the already existing gender gap if REDD and REDD+ regimes are not designed carefully.

Despite these facts, women have not been afforded an equal opportunity to participate in decision-making related to adaptation and mitigation policies and initiatives at the international and national level related to climate change.

Strategic steps

Based on these observations and analyses the following strategic steps are suggested.

Improve the understanding of gender and climate change

It is necessary to integrate the social, economic, political and cultural dimensions in the examination of the causes and consequences of climate change. Particularly an analysis of gender relations should be included. It is also urgent to develop interventions with a gender perspective for the different scenarios related to climate change and to design strategies that can help to make gender relations more equitable.

Develop gender provisions in international policy instruments

It is important to develop effective international and national legal instruments, so that they can contribute as guiding principles that could mitigate the effects and promote the adaptation to climate change more equitably among human beings.

The international negotiation process within the UN Framework Convention on Climate Change (UNFCCC), as well as the regional, national and local policies on climate change should incorporate the principles of gender equity and equality throughout all its stages: from research to analysis, from design to implementation of mitigation and adaptation strategies. Particularly, these should be applicable in the process towards a system or 'regime' for the protection of the climate after the year 2012.

Recognition of the importance of gender in meetings of the UNFCCC is of the highest importance. It is necessary that state parties take the necessary measures for the UNFCCC to abide by the human rights frameworks and international and national frameworks regarding gender, with special attention to the Convention on the Elimination of all Forms of Discrimination Against Women (CEDAW) and the recommendations of the Experts Committee of the CEDAW (see also Chapter 7).

The UNFCCC needs to develop a gender strategy and invest in specialized research on gender and climate change (for example: patterns of use and specific resources by gender; climate change effects by gender; gender aspects of mitigation and adaptation; women's capacity to cope with climate change, vulnerability and its gender patterns, among others), and establish a system of gender-sensitive indicators for the national reports presented to the Secretariat of the UNFCCC, for the planning of the adaptation strategies or projects under the Clean Development Mechanism (CDM) or REDD. For example, as the importance of land and resources to women is frequently neglected in national statutory and customary laws, REDD projects should comply with international agreements related to women's equal access to land ownership and resources.

State parties should seek for mechanisms that could guarantee the participation of women and gender experts during the preparation of national and international reports. Likewise, they should ensure women's representation in the national and international meetings of the UNFCCC.

More specific information on legal frameworks on gender and climate change is presented in Chapter 7.

Mainstream gender in national and local political actions

Governments and local authorities have a crucial role in guiding implementation and delivery. Therefore gender should be mainstreamed into all national policies, action plans and specific projects on climate change. This can be achieved through conducting systematic gender analysis, creating gender indicators, clear targets on inclusion of women and developing tools for capacity building on gender issues.

Specific actions for national and local governments are:

- Translate international agreements on gender and climate change (treaties, agreements, conferences, declarations and resolutions) to their internal policy.
- Develop strategies to improve and guarantee women's access to and control over resources.
- Use the specialized knowledge and skills of women in the strategies for survival and adaptation to disasters.
- Create opportunities for the education and training of women on climate change.
- Provide measures for capacity building and technology transfer.
- Assign specific resources to secure women's equal participation of the benefits and opportunities of the mitigation and adaptation measures.
- Mainstream gender in design, implementation and assessment of projects.

Encourage gender-sensitive financial mechanisms and instruments

Gender analysis of all financial instruments to address climate change is essential to ensure gender-sensitive channelling of resources for adaptation, mitigation, technology transfer and capacity building. Such financing mechanisms should reflect women's priorities and needs.

All stages and aspects related to the financial mechanisms and instruments associated to climate change should include, in their principles, the mainstreaming of a gender perspective and women's empowerment. Some of the essential areas are the design, implementation and monitoring and evaluation systems. This can be achieved by including gender-sensitive criteria into the development of programmes, projects or initiatives that are under the financial mechanisms.

Some examples of how this can be achieved are:

- Adaptation funds should guarantee the incorporation of gender considerations and the implementation of initiatives that could satisfy women's needs. A gender appraisal should be mandatory for all projects aiming for financial support.
- Women should be included in all levels of the design, implementation and evaluation of the projects of forestation, reforestation and conservation that receive payments for environmental services, such as carbon sinks.
- Women should be able to access commercial carbon funds, credits and information that can enable them to understand and decide which modern resources of biomass and technologies meet their needs.
- The CDM should finance projects that bring renewable energy technologies within the reach of women and that could supply their domestic needs (see Box 6.3).
- A gender perspective should be included in existing international standards and guidelines on REDD to ensure that women have equal access to and control of all REDD benefits, as well as gender responsive indicators (see Table 6.2).
- In the implementation of REDD schemes, special attention should be given to the gender implications of benefit sharing and payment structuring frameworks and to including gender responsive indicators and targets as part of performance-based funding.

Involve women in technological developments

Women's needs, priorities, technological needs, knowledge and traditional practices have to be taken into account in developing new technologies. Ensuring direct involvement of women in technology development can help guarantee ownership by users, effectiveness and sustainability. Gender mainstreaming has to be included in creating enabling environments, capacity building, and mechanisms for technology transfers to ensure full participation of both men and women in climate change technological initiatives (Karlsson, 2007).

Promote active participation of women's organizations

The organizations, ministries or departments of women's issues and related UN entities specializing in gender should have a more active role in the discussions and decisions that are being made on climate change. Climate change cannot be considered exclusively an environmental problem. It should rather be understood within all its development dimensions.

Box 6.3 *Nepal National Biogas Project: Reducing emissions while providing community benefits*

Only about 10–15 per cent of people in the rural areas of Nepal have access to electricity. The Nepal Biogas Project promotes the use of biogas in Nepal for cooking and lighting in rural households by offering biogas units at below market cost. The project activities reduce GHG emissions by replacing current fuel sources (mostly firewood, dung and kerosene) with biogas produced from animal and human wastes.

In households with biogas units, women benefit through reduced time and effort in collecting and managing fuelwood supplies. They are also exposed to fewer of the health risks associated with indoor air pollution from smoky fires and kerosene lamps. The project estimates that women save three hours daily per household when they use biogas for cooking rather than collecting firewood. Women report that they use the saved time in income-generating efforts, attending literacy classes, social work and recreation.

When households directly connect their latrines to biogas production units, they and their communities enjoy better health and sanitation. In addition, there are new employment opportunities connected with biogas digester production and distribution.

This was the first GHG emission reduction project in Nepal approved for financing under the CDM. Developed by the Alternative Energy Promotion Centre, the project is obtaining financing for subsidized distribution of the biogas units by selling a total of 1 million tons of GHG emission reductions to the Community Development Carbon Fund managed by the World Bank. The project estimates that each household biogas unit will eliminate close to five tons of carbon dioxide equivalent per year. Selling emissions reduction credits allows the project to generate long-term funding without seeking ongoing assistance from donors.

The country's dependence on fuelwood has contributed greatly to deforestation, so the project will also reduce pressures on the forests. In addition, waste slurry from the biogas digesters can be used as organic fertilizer, boosting food production and avoiding the expense of buying chemical fertilizers.

Sources: World Bank Carbon Finance website, http://carbonfinance.org/Router.cfm?Page=CDCF&FID=9709&ItemID=9709&ft=Projects&ProjID=9596; Forestry Nepal, *Nepal Biogas Project: Reducing emissions while providing community benefits*, www.forestrynepal.org/article/news/375, accessed 25 February 2010

Involving women's organizations in the decision-making process on climate change will bring much needed expertise on women's rights and gender considerations.

Local women's organizations should be recognized as stakeholders and constituents in policy development and planning and given special provisions for active participation, including having a speaking role and being listened to. Promoting the involvement of women's groups may be a challenge, because they are not homogenous but organized among different social, minority, age and economic groups. However, by actively promoting participation of local women's groups, the capacity of women can be strengthened, leading to the empowerment of women, public engagement, respect for their communities, and opportunities for women to assume leadership roles. This can also enhance their access to additional income and economic opportunities which are all essential for women's advancement. Additionally, involvement of women in climate change initiatives is important, because women in many countries are responsible for transmitting education and environmental principles to children.

The promotion of women's participation in climate change projects should be systematic and comprehensive. For example, with regard to REDD, women's groups should participate in REDD consultations, national REDD working groups, and the design, implementation, monitoring and evaluation of REDD projects. Women groups should have equal and timely access to information regarding REDD planning and implementation, and access to capacity-building opportunities, to enable full and effective participation. REDD schemes and research should value women's traditional and scientific knowledge and their entrepreneurial potential in relation to combating deforestation and degradation of forests (see also Table 6.2).

Mainstream gender in climate change policies and actions

For the climate change regime, strategies and projects to be sustainable and efficient, gender has to be mainstreamed in international policy instruments, national and local political actions, financial mechanisms and instruments, and technological developments. Women have to be provided with equal and equitable opportunities to participate in decision-making related to adaptation and mitigation policies and initiatives at the international and national levels. Mainstreaming gender and increasing participation of women should be included comprehensively and systematically in climate change policy development, implementation and evaluation phases.

In Table 6.2, an example is presented of comprehensive gender mainstreaming in climate change mitigation measures and financial mechanisms; in this case a REDD Readiness Plan.

Table 6.2 *Example of a gender mainstreaming process: REDD*

ENGENDERING A REDD READINESS PLAN	
Readiness plan (R-Plan)	*Gender consideration*
1. Land use, forest policy and governance quick assessment: prepare an early analytic assessment of past experience to reduce deforestation, to identify promising approaches and lessons learned. Analyze governance and legal issues related to land use pertinent to REDD actions. Independent or external expert authors may be commissioned to enhance objectivity and quick preparation.	Expanding the scope of the assessment to include: Land use: • Gain basic socio-economic and cultural information disaggregated by sex on communities within the project zone and nearby communities affected by the project. • Understand the current land use and forest management practices through data disaggregated by sex; including different uses of the forest by men and women, their respective roles in deforestation and degradation of forests. • Understand land rights issues; including customary and legal property rights concerning women's access to land as well as international agreements applicable in the host country. Forest policy: • Analyse forest policy of national, regional, sub-regional and local levels as appropriate to verify whether gender is well incorporated therein. Governance: • Understand which legal and customary institutions are responsible for land use and REDD-related decisions. • Verify whether within legal and customary institutions gender balance is ensured and women's interests are addressed. • Acknowledge which legal and customary institutions' approval is needed for REDD projects ensuring equitable access and sharing of benefits for women. • Identify and support work that women groups are doing in the governance of forests.
2. Management of readiness: 2a. Convene a National REDD Working Group: Present the design of a national working group to coordinate Readiness activities and ultimately REDD implementation, its methods of operation, and how REDD will be integrated into the existing land use policy dialogue. The working group process should include internal and external stakeholders, and the coordination of donor efforts supporting REDD or land use activities.	• Include in the working groups at national level a gender and forestry expert. • Include women as separate stakeholder group into the consultation process and address their capacity-building needs for participating. • Invite women's organizations to the working group process.

Table 6.2 *Example of a gender mainstreaming process: REDD (continued)*

ENGENDERING A REDD READINESS PLAN

Readiness plan (R-Plan)	Gender consideration

2b. Prepare a REDD Consultation and Outreach Plan:
Prepare a REDD consultation and outreach plan, to ensure continuous, inclusive consultation during the development (and eventual implementation) of the REDD strategy, implementation framework, reference scenario, monitoring system, and other R-Plan components during the Readiness phases. Special attention should be given to discussion and assessment of potential social and environmental impacts of the evolving REDD strategy.
	• Include the promotion of gender equality as goal into the design and implementation of both REDD Consultation and Outreach Plan.
	• Consult women and women's groups during the design and implementation of consultation processes (gender-sensitive consultation methods).
	• Address the following issues during the Consultation processes and in the Outreach Plan:
	  – Identify positive and negative socio-economic impacts differentiated for women and men,
	  – Identify risks (social, economic and cultural factors) to the equitable distribution of socio-economic benefits between women and men.
	  – Understand the added value of gender mainstreaming through comparing scenarios 'with gender' and 'without gender'.

3. Design the REDD strategy:
3a. Assess candidate activities for a REDD Strategy:
Summarize the outlines of a REDD strategy and candidate activities, building on the land use policy assessment (above), stakeholder consultations, and analytic work. Summarize the knowledge and capacity gaps, and analytic activities needed to elaborate and define a REDD strategy.
	Main elements of REDD strategy and candidate activities:
	• Include the promotion of gender equality as one of the goals of the strategy.
	• Identify the potential risks for equitable benefits for women deriving from the project.
	• Describe each project activity in relation to the promotion of gender equality.
	• Identify gaps in capacity and knowledge:
	  – Raise the awareness on the importance of gender among project staff and stakeholders.
	  – Enhance the capacity of women and women's groups to participate in the project.
	• Identify clearly who is responsible for gender mainstreaming among members of project staff during both strategy development and implementation.
	• Show how financial mechanisms sponsoring the REDD activity ensure the necessary flow of funds for gender-related activities and equitable distribution of benefits.
	• Develop a monitoring and evaluation system which includes variables on gender.

3b. Evaluate potential additional benefits of REDD, including biodiversity conservation and rural livelihood: Conduct an assessment of potential benefits of the REDD strategy for	• Analyse by sex the beneficiaries of current distribution mechanisms of forest benefits, at national and local level. • Conduct a gender-sensitive assessment of potential benefits of REDD (identify different benefits for men and women).

Table 6.2 *(continued)*

ENGENDERING A REDD READINESS PLAN

Readiness plan (R-Plan)	Gender consideration
biodiversity conservation and rural livelihood, and other benefits deemed important by a country (for example, water supply).	• Identify possible negative impacts on women of the REDD Strategy and identify actions to mitigate those. • Find solutions for equitable benefit sharing between women and men.
3c. Trade-offs analysis: Assess the trade-offs across candidate elements of your REDD strategy in terms of your broader land use policy dialogue and sustainable development policies, to help define an integrated REDD strategy.	• Gender equality (aspects) should be considered as a component of the decision-making process for defining the trade-offs on REDD. • Access how trade-offs (for competing land uses for example) affect women's livelihoods strategies and practices.
3d. Risk assessment of your REDD strategy: Evaluate barriers to successful implementation of the REDD strategy, risks associated with the proposed strategy elements, and ways to reduce or compensate for those risks.	• Identify how the lack of incorporating gender into REDD Strategy might impact effectiveness and sustainability (lack of stakeholder commitment, loss of knowledge, increased poverty, health and nutrition impacts etc).
4. REDD implementation framework: Assess the institutional arrangements and legal requirements needed to implement REDD activities, including design of an equitable payment mechanism. Issues to be analysed and addressed are likely to include ownership of carbon rights, equitable revenue sharing mechanisms, national carbon registry to manage different REDD activities and revenue streams.	• Assess formal and customary laws having impact on women's participation in REDD strategies and the applicable international agreements in the host country. • Evaluate the appropriate manner for the executing institutions to mainstream gender (gender policy, focal point, criteria standards, gender plan of action, among others). • Assess institutional gaps and opportunities at the local and national levels for women's groups to engage in REDD and the private carbon markets.
5. Assess the social and environmental impacts of candidate REDD strategy activities: Assess potential impacts by performing an impacts assessment, using the Environmental Strategic Management Framework or another analytic approach. Feed this assessment into the consultation plan and ongoing consultations.	• Carry out a gender-sensitive impact assessment through comparing the scenarios 'with REDD strategy' and 'without REDD strategy' considering the impact separately for both women and men. • The impact assessment should be carried out with gender-responsive participatory methods.
6. Assess investment and capacity building requirements: Assess candidate REDD strategy elements and the REDD implementation framework, in terms	• Assess the needs for capacity building of women and women's groups. • Verify whether the financial resources available can support activities carried out by women within the REDD strategy.

Table 6.2 Example of a gender mainstreaming process: REDD (continued)

| ENGENDERING A REDD READINESS PLAN | |
Readiness plan (R-Plan)	Gender consideration
of capacity requirements, financial support needed, and gaps existing with regard to potentially available resources.	
7. Develop a reference scenario of deforestation and degradation: Develop objectives; a work plan to realize those objectives during the R-Plan implementation phase; and prepare the terms of reference for the majority of that work plan.	• Include gender specific objectives in the reference scenario. • Include the need for gender expertise in the terms of reference.
8. Design and implement a monitoring, reporting and verification (MRV) system for REDD: Provide the capacity to monitor forest sector carbon emissions and other characteristics over time, in relation to the Reference Scenario.	• Identify gender-related variables. • Consider the possibility of establishing participatory monitoring systems including both women and men. • Ensure that the monitoring and evaluation are based on sex-disaggregated data.
9. Design a system of management, implementation, and evaluation of Readiness preparation activities (optional): Synthesize all R-Plan components into a REDD national programme that is effectively and transparently managed, and regularly evaluated using pre-established indicators of performance and effects on development plans.	• Involve women, groups of women and women's organizations into the entire readiness process, including in all its economic, social, technical and financial elements.

Developed by L. Aguilar, C. Espinoza, A. Quesada-Aguilar, R. Pearl and A. Sasvari (2009); based on World Bank (2009) and CCBA (2008).

Notes

1   Differentiated data on access of women and men to assets can be found in different institutions in the UN system, such as the UNDP's Gender-related Development Index (GDI) (UNDP, 1995), UN Development Fund for Women (UNIFEM), International Research and Training Institute for the Advancement of Women (INSTRAW), Food and Agriculture Organization (FAO) and International Fund for Agricultural Development (IFAD).

2   REDD+: reducing emissions from deforestation and forest degradation in developing countries; and the role of conservation, sustainable management of forests and enhancement of forest carbon stocks in developing countries. This is based on paragraph 1 (b)(iii) of the Bali Action Plan, which was adopted at the 13th session of the Conference of Parties (COP13) in December 2007.

The main difference between REDD and REDD+ is that REDD+ gives the same level of priority to conservation, sustainable management and the enhancement of forest carbon stocks as to deforestation and degradation of forests.

References

Agarwal, B. (2002) 'Participatory exclusions, community forestry, and gender: An analysis for South Asia and a conceptual framework', *World Development*, vol 29, no 10, pp1623–1648

Aguilar, L., Araujo, A. and Quesada-Aguilar, A. (2007) *Reforestation, Afforestation, Deforestation, Climate Change and Gender*, factsheet, International Union for Conservation of Nature, Costa Rica

Anker, R. (1998) *Gender and Jobs: Sex Segregation of Occupations in the World*, International Labour Organization, Geneva

CCBA (Climate, Community and Biodiversity Alliance) (2008) *Climate, Community and Biodiversity Project Design Standards*, 2nd Edition, CCBA, Arlington, VA, www.climate-standards.org, accessed 24 February 2010

Drexhage, J. (2006) *The World Conservation Union (IUCN) Climate Change Situation Analysis: Final Report*, International Institute for Sustainable Development (IISD)/ IUCN, Winnipeg/Gland

Donner, S. D., Knutson, T. R. and Oppenheimer, M. (2007) 'Model-based assessment of the role of human-induced climate change in the 2005 Caribbean coral bleaching event', *Proceedings of the National Academy of Sciences*, vol 104, no 13, pp5483–5488

Karlsson, G. (ed) (2007) *Where Energy is Women's Business: National and Regional Reports from Africa, Asia, Latin America and the Pacific*, ENERGIA, Leusden

FAO (Food and Agriculture Organization of the UN) (2001) *Women – Users, preservers and managers of agrobiodiversity*, factsheet, FAO, Rome, www.fao.org/ sd/2001/PE1201a_en.htm, accessed 24 February 2010

FAO (2006) *Building on Gender, Agrobiodiversity and Local Knowledge: A Training Manual*, FAO, Rome

FAO (2008) 'Gender and food security: Agriculture', Factsheet, FAO, Rome

Green Belt Movement (2009) *Responding to Climate Change from the Grassroots: The Green Belt Movement Approach*, Green Belt Movement, Nairobi

Hope to Action (2010) 'Women for a greener planet', www.hopetoaction.org, accessed 24 February 2010

Inter-American Development Bank (1999) *Hurricane Mitch: Women's Needs and Contributions*, Technical Papers Series, Inter-American Development Bank, Sustainable Development Department, Washington

IPCC (Intergovernmental Panel on Climate Change) (2001) *Summary for Policymakers. Climate Change 2001: Impacts, Adaptation, and Vulnerability*, report of IPCC Working Group 2 (WG2), IPCC, Geneva

IPCC (2007) *IPCC Fourth Assessment Report: Climate Change 2007*, IPCC, Geneva

IPCC (2008) *Climate Change and Water*, technical paper of the IPCC, IPCC Secretariat, Geneva

IUCN (International Union for Conservation of Nature) (2009) *The IUCN Red List*, IUCN UK Office, Cambridge, www.iucnredlist.org, accessed 25 February 2010

Kenya NBSAP (National Biodiversity Strategies and Action Plan) (2000) www.cbd.int/nbsap/search, accessed 24 February 2010

Liberia NBSAP (2004) www.cbd.int/nbsap/search, accessed 24 February 2010

Macchi, M., Oviedo, G., Gotheil, S., Cross, K., Boedhihartono, A., Wolfangel, C. and Howell, M. (2008) *Indigenous and Traditional Peoples and Climate Change*, issues paper, IUCN, Gland

Moser, C. and Satterthwaite, D. (2008) *Pre-poor Climate Change Adaptation in the Urban Centres of Low-and Middle-Income Countries*, Workshop on Social Dimension of Climate Change, World Bank, Washington, DC

Nelson, V., Meadows, K., Cannon, T. and Morton, J. (2002) 'Uncertain predictions, invisible impacts, and the need to mainstream gender in climate change adaptations', *Gender and Development*, vol 10, no 2, pp51–59

Neumayer, E. and Plümper, T. (2007) *The Gendered Nature of Natural Disasters: The Impact of Catastrophic Events on the Gender Gap in Life Expectancy, 1981–2002*, London School of Economics, University of Essex and Max Planck Institute for Economics, London

NFWI and WEN (National Federation of Women's Institutes and Women's Environmental Network) (2007) 'Women's manifesto on climate change', NFWI and WEN, London, www.wen.org.uk/wp-content/uploads/manifesto.pdf, accessed 24 February 2010

OECD (Organisation for Economic Co-operation and Development) (2008a) *Promoting Sustainable Consumption: Good Practices in OECD Countries*, OECD, Paris

OECD (2008b) *Environmental Policy and Household Behaviour: Evidence in the Areas of Energy, Food Transport, Waste and Water*, OECD, Paris

Oxfam (2006) *Drought Management Consideration for Climate Change Adaptations: Focus on the Mekong region*, Oxfam in Vietnam and Graduate School of Global Environmental Studies of Kyoto University, Hanoi and Kyoto

REDD (2009) 'About REDD', www.un-redd.org/AboutREDD/tabid/582/language/en-US/Default.aspx, accessed 24 February 2010

Rodó, X., Pascual, M., Fuchs, G. and Faruque, A. S. G. (2002) 'ENSO and cholera: A nonstationary link related to climate change?', *Proceedings of the National Academy of Sciences*, vol 99, no 20, pp12901–12906

The Equilibrium Fund (2007) 'Reforestation, food forests and carbon offsets', www.theequilibriumfund.org/page.cfm?pageid=5494, accessed 24 February 2010

UNDP (United Nations Development Programme) (1995) *Human Development Report 1995*, UNDP, New York

UNEP (United Nations Environment Programme) (2010) 'The billion tree campaign: Growing Green', www.unep.org/billiontreecampaign/index.asp, accessed 24 February 2010

van Cotthem, W. (2009) 'Women and climate change: Issues of impact, equity, and adaptation', http://desertification.wordpress.com/2009/01/18/women-and-climate-change-issues-of-impact-equity-and-adaptation-google-celsias, accessed 24 February 2010

WHO (World Health Organization) (2009a) *Gender, Health and Climate Change*, Draft Discussion Paper, WHO, Geneva

WHO (2009b) *Pregnant Women and Infants*, Global Malaria Program, WHO, Geneva

World Bank (2009) *Readiness Mechanism and Readiness Plan Criteria*, World Bank, Washington, DC

World Bank, IFAD and FAO (2009) Gender in Agriculture Sourcebook, World Bank, Washington, DC

Zhou, G., Mina Kawa, N., Githoko, A. and Yan, G. (2004) 'Association between climate variability and malaria epidemics in the East African highlands', Proceedings of the National Academy of Sciences, vol 101, no 8, pp2375–2380

7

Climate Change and Gender: Policies in Place

Tracy Raczek, Eleanor Blomstrom and Cate Owren

Introduction: The roots of international policy and norms related to gender and climate change

International norms and policies specifically related to gender and climate change have been slow to emerge. However they are increasingly surfacing, carving out space in the nexus between two more long-standing regimes – the environmental regime and the human rights regime. Principles expressed in the international agreements of these two arenas currently provide the foundation, and in some cases specific language, from which principles and policies have been drawn to address the gender dimensions of climate change. This chapter will examine this pattern of norm-setting on the issue in the international arena, attempting to take a snapshot of its current status amid the ongoing climate change negotiations and persistent issue of gender inequality, both of which may define this century's international development agenda.

For various reasons, no single international agreement encompasses all the components of climate change (energy, adaptation and disaster risk reduction, environmental degradation, exacerbated poverty, human rights, and so on), thus a combination of principles outlined in many agreements and instruments must be compiled to fully complete the climate change picture. Among the many multilateral environmental agreements (MEAs), some assail the Kyoto Protocol as 'the' climate change agreement, but it falls short of this broad scope or mandate and its successor has yet to be adopted. Similarly, while the Convention on the Elimination of All Forms of Discrimination against Women (CEDAW) is assailed as the women's bill of rights, multiple instruments of the human rights regime must be drawn upon to locate references linking women's rights and other gender equality principles to climate change impacts and responses – such as economic and livelihood rights, access to resources, equitable distribution of benefits and opportunities and participation in decision-making.

Both climate change and gender equality are cross-cutting issues. This provides an advantage in that references to both fall into multiple instruments, providing more norm-setting opportunities. For example, cornerstone environmental instruments that moderately or very deliberately incorporate gender equality principles include the UN's Agenda 21, Convention to Combat Desertification (UNCCD) and Convention on Biological Diversity (UNCBD). Simultaneously, cornerstone gender equality instruments that include references to environmental and sustainable development issues include the 1976 CEDAW and the more recent Beijing Declaration and Platform for Action of the late 1990s. While references found in all of these are relevant to the climate change agenda, their breadth demands synthesizing to extract whether and how they can be drawn upon to forge principles that empower women facing global warming, in particular. Other instruments, less well-known but as critical including the Hyogo Framework for Action and the UN Economic and Social Council (ECOSOC) Resolution 2005/31 on gender mainstreaming in the UN, also deserve attention for being mainstays of gender mainstreaming in key climate change areas.

This chapter provides an outline of the main international instruments currently used to construct the gender and climate change regime, and an overview of where they fall on the spectrum of 'soft' to 'hard' legal instruments. Additional text boxes detail highlights of the gender mainstreaming history in the environmental and human rights regimes and elucidate the striking moments where it has claimed its own space. The scope of this chapter does not include an in-depth analysis of the socio-political or economic framework in which these instruments are negotiated, nor the influence of related movements such as neo-liberalism or eco-feminism. It is intended to provide a concise reflection of the international instruments – both conventional and customary — which provide the foundation for nascent but critical policies and norms intended to address the gender dimensions of climate change. Notably these instruments are increasingly drawn upon by policy-makers, judicial bodies and civil societies to inform policies and practices at the regional, national and local levels.

Box 7.1 *Environmental and women's rights movement of the
1960s and 1970s*

The late 1960s through to the late 1970s was a decade of revolutions that catapulted the environmental and women's rights movements to new heights. The environmental movement of this specific era was inspired by views of the planet Earth from space and backed by new scientific understandings of ecological interdependence and human's impact on the Earth and their own health. It was promoted by civil society, media, politicians and economists that dared policy-makers and the private sector to incorporate externalities. The movement gave birth to a multitude of non-governmental organizations (NGOs) such as Greenpeace (founded 1971) as well as national and international efforts to institutionalize environmentally sound policies, including the US Environmental Protection Agency (EPA, founded 1970) and the UN Environmental Programme (UNEP, founded 1972).

The environmental movement was matched in scale and significance by a solid push for women's rights. Worth noting this movement grew, at least in the US, in great part from advances made by and synergies with the civil rights movement advancing racial equality, the Civil Rights Act of 1964 and a rich history of efforts by earlier women's rights activists. That said, unique opportunities and advances of this decade afforded many women choices as never before, including unprecedented access to contraception and family planning. Women joined universities and non-traditional professions in record numbers. Civil society organized (for example, the renowned National Organization for Women was founded in 1966) and advocates pushed for equal rights and an end to discrimination on many fronts. The US Congress responded with, among other things, the Equal Rights Amendment (passed 1972) and Title IX in the Education Codes (passed 1972), which further enabled a social revolution and surge of women in the workplace and politics. While both the environmental and women's rights movements can be viewed as second, third or even fourth waves of earlier movements also aimed to advance their causes – such as the women's rights movement of the late 1800s to the early 20th century or the wilderness conservation movement of the late 19th century – this particular decade was significant in that it succeeded in laying much of the institutional framework for contemporary policy-making and fed into what are increasingly considered core cultural norms, not only in the US but worldwide.

Hard versus soft: Conventional versus customary instruments

Multinational intent to address environmental and human rights challenges, especially those with transboundary causes and ramifications, is increasingly codified in international instruments. This has led to a profusion of agreements, conventions, protocols and declarations since the 1969 Vienna Convention of the Law of Treaties, which negotiated the rules and procedures for such a process, and most notably since it entered into force in 1980.

Of these, the instruments included in the environmental and human rights regimes are primarily regarded as customary or 'soft' law, as are the majority of all international agreements. This reflects their historically lower status in the international arena and the specific fora in which they are negotiated for reasons too complicated to indulge here. In the end, they are significantly less robust with fewer mechanisms for compliance than conventional law, especially compared to instruments that address traditional security threats, such as high crimes and war. In the human rights regime, however, crimes against humanity are the main exception as the International Criminal Court can be called upon to address impunity when other routes are exhausted. Nonetheless, while soft instruments are often criticized as ineffectual, they offer many advantages. They can be arrived at more quickly especially when consensus is improbable, and they can reflect and influence norms of non-state actors as well as states. Thus they play a critical role in long-term international and national customs, norms and practices on the ground, especially when used as advocacy tools by civil society and domesticated by states into national laws.

Soft and hard instruments vary in their leverage, scope and durability, and while it is difficult to generalize, patterns exist. Leverage, for example, tends to stem from the forum in which agreements were born as well as the content of the agreement. Within the UN, for example, the leverage of a resolution is generally higher when sourced to the Security Council, and decreases when sourced to the General Assembly, followed by additional subsidiary UN bodies. The scope, which includes details of its intended sphere of influence and the delegation of authority, also varies widely. A resolution, for example, requesting that further studies be completed falls short of a resolution requesting action by all parties to the agreement; and preambular text 'noting' actions by states falls short of a resolution 'requesting' or 'urging' action. Details matter. Durability aligns with both these circumstances, as some resolutions are intended for perpetuity while others are reconsidered regularly, even annually. These qualities as well as political context and the more obtuse consideration of authors' intent also contribute to affect the weight of any given instrument.

Therefore, not much can be gleaned from the title of an instrument. No absolute nomenclature exists for terms assigned to instruments because titles can differ substantially, even among instruments of relatively equal leverage and stature. That said, there is a general spectrum onto which instruments fall from more robust 'treaties' down through softer 'declarations' and 'plans of action', and the title of an instrument can be used as a first-step, although simplistic, assessment tool to understand the weight of an instrument.

In short, the term 'treaty' has been used to designate agreements that are drafted by parties with treaty-making power, are legally binding and commonly require ratification. This term is often reserved for issues of gravity; however, it is used increasingly less often, replaced by other titles such as 'convention'. The term 'convention' is typically recognized as having equitable legal weight to 'treaty'. However, a convention is often negotiated in more inclusive international fora under the auspices, for example, of an international organization or conference such as the UN Conference on Environment and Development (UNCED) that allows input from stakeholders other than states, and therefore is considered by some to have less leverage than a purely state-centric negotiation process. The term 'declaration' is often the most difficult to decipher as context is critical. While the term can often denote a softer instrument, not intended to be legally binding, states can also use a declaration – unilaterally or otherwise – to denote an arrangement with legal implications or as an annex to a legally binding instrument. Declarations may also reflect well-established customary law and have increased leverage in these circumstances as is the case with the UN Universal Declaration of Human Rights (UN-UDHR). The term 'agreement' is a catch-all phrase, often interchangeable with 'instrument' although not as broad or inclusive (UN, 2008). Overall, the broad array of instruments is best understood by examining the circumstance or formality under which each agreement was formed. Figure 7.1 represents the wide spectrum on which hard and soft instruments can fall.

Development and status of relevant international instruments

Table 7.1, below, provides an overview of the various instruments that include principles and norms applicable to the gender and climate change

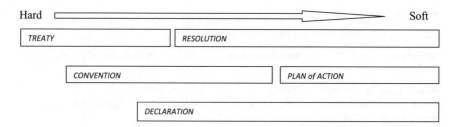

Figure 7.1 *Spectrum of hard and soft international instruments*

Source: T. Raczek

agenda. Worth noting, the process to develop international agreements can take years, even decades. The fora and dates provided below represent the time and place they were opened for adoption or fully adopted, and do not reflect the sometimes multitude of years and fora needed for the development, introduction, adoption and/or ratification process. The final column provides a general assessment of the instruments' gender perspective, rating them as having strong, moderate, limited or an absence of gender perspective. An instrument with strong gender perspective may advance gender mainstreaming in all aspects of its implementation, for example, or have multiple references to gender-specific dimensions as well as advocate for women's participation in all areas of policy or programming – from design through governance. Instruments with moderate gender perspective may reflect the latter – participation of women at many levels – or include a few strong development or rights objectives that specifically target women. Instruments with limited gender perspective are those that have few and/or weak objectives or mechanisms that specifically aim to empower women. For example, the instrument may refer to ensuring the rights of women are 'paid attention to', with little or no other reference to gender specific issues or mainstreaming of gender. Instruments absent of gender perspective are those with no reference specifically to women or gender dimensions. Whether this is intentional or the authors considered the issue to be free of gender dimensions (gender neutral) is not assessed here. However the causes for an absence of gender perspective may differ significantly and context should be considered in all cases. For example it is often presumed that the 1948 UN-UDHR has limited gender specific language for very different reasons than the 1992 UN Framework Convention on Climate Change (UNFCCC).

Table 7.1 *Key instruments that include principles and norms applicable to the gender equality and climate change agenda*

Instrument	Forum and date	Assessment of gender perspective
Human rights and gender equality		
UN Universal Declaration of Human Rights (UN-UDHR)	UN General Assembly; 1948	Limited gender perspective
Mexico Declaration and Plans of Actions	1st World Conference on Women; 1975	Strong gender perspective
Convention on the Elimination of All Forms of Discrimination against Women (CEDAW)	Adopted by the UN General Assembly; 1979	Strong gender perspective
Beijing Declaration and Platform for Action	4th World Conference on Women; 1995	Strong gender perspective
United Nations Economic and Social Council (ECOSOC) Resolution E/1997/66	Adopted by ECOSOC, 1997; Numerous subsequent resolutions, including 2005/31 on gender mainstreaming in the UN	Strong gender perspective
United Nations Declaration on the Rights of Indigenous Peoples (DECRIPS) A/RES/61/295	Adopted by the UN General Assembly; 2007	Limited gender perspective
Environment		
Declaration of the UN Conference on the Human Environment	UN Conference on the Human Environment; (Stockholm); 1972	Absence of gender perspective
Agenda 21 and the Rio Declaration on Environment and Development	UN Conference on Environment and Development (UNCED); 1992	Strong gender perspective
UN Convention on Biological Diversity (UNCBD)	UNCED; 1992	Moderate gender perspective
UN Convention to Combat Desertification (UNCCD) and related General Assembly Resolution 47/188	UNCED; 1992, and General Assembly; 1994	Moderate gender perspective
Sustainable development and disaster risk reduction		
Millennium Declaration and Millennium Development Goals (MDGs)	Adopted by the UN General Assembly; 2000 and 2001	Moderate gender perspective
Hyogo Framework for Action (HFA)	World Conference on Disaster Reduction; 2005	Strong gender perspective
Climate change		
United Nations Framework Convention on Climate Change (UNFCCC)	UNCED; 1992	Absence of gender perspective
Human Rights and Climate Change Resolutions 7/23 and 10/4	Adopted by the Human Rights Council; 2008 and 2009 respectively	Moderate gender perspective

International instruments and their applicability to gender and climate change agenda

World leaders from every corner of the globe have agreed upon a multitude of international conventions, resolutions, declarations, frameworks and platforms for action with reference to either environmental or human rights issues, as well as more specifically to climate change or gender equality issues. Together these agreements are raising international awareness about the gender dimensions of climate change and add to the framework for climate change and gender equality. In some cases, the agreements indicate specific government commitments and areas of priority for action. The following section will elaborate on these agreements in order from 'hardest' to 'softest.'

Conventions and Resolutions

Conventions

Several UN Conventions inform the framework for gender and climate change policy. The most comprehensive global agreement on advancing gender equality, CEDAW, was adopted by the UN on 18 December 1979. Commonly known as the first international bill of rights for women, it addresses women's human rights for land ownership, resources, livelihoods, education and safety, and is relevant for climate change policies and programmes in numerous ways. CEDAW obliges parties to take all necessary measures to ensure women benefit from rural development and are involved in all aspects of planning for development. In particular, Article 11 calls for equal employment opportunities including vocational

Box 7.2 *Early environmental international treaties*

The challenge of addressing environment and development issues across national borders is not unique to the last few decades. Members of the global community have long struggled with transboundary issues and have collaborated to mitigate their damage. In 1878, for example, the first international treaty dealing with trade and the environment, the Phylloxera agreement, restricted the trade of grapevines in order to prevent a dangerous spread of pests attacking vineyards (Harris, 2006). Numerous international agreements since then have dealt with environmental issues ranging from conservation of wildlife to limiting hazardous wastes to preserving clean waters.

training; Article 14 calls for equal access to agricultural credit and loans, marketing facilities and appropriate technology and equal treatment in land reform and resettlement schemes (CEDAW, 1979). When applied to the climate change agenda these Articles may facilitate adaptation and enhance women's resilience to impacts of climate change. Moreover, in a specific link to climate change in 2009, members of the CEDAW Committee adopted climate change as an urgent issue of focus, recognizing the gender differentiated impacts of climate change and calling on parties to include gender equality as a guiding principle when drafting the new international climate change agreement expected to extend or replace the Kyoto Protocol post-2012 (CEDAW, 2009).

A major moment in international environmental policy history – as well as for gender equality in policy-making – came in 1992 at the UNCED, also referred to as the Rio Conference and most recognized as the Earth Summit, in Rio de Janeiro, Brazil. The Earth Summit produced five outcome documents: the Rio Declaration and Agenda 21, as well as three major environmental conventions. Agenda 21, although not legally binding, is the first and, to date, the only global blueprint for sustainable development. It is one of the first major UN documents to comprehensively incorporate women's roles, positions, needs and expertise throughout. The document references women in terms of, among other things, outreach, training, participation and decision-making, health, land management, water resources and the need for gender-disaggregated data and gender-specific programme evaluation. Notably, Chapter 24 of Agenda 21, 'Global Action for Women towards Sustainable and Equitable Development', identifies areas that require urgent international action to achieve equality between women and men, which in turn is necessary to enable effective implementation of the sustainable development agenda (UNCED, 1992).

Moreover, two of the three major UN Conventions linked to UNCED articulated gender equality principles. Signed by 150 governments at the Earth Summit, the UN Convention on Biological Diversity (UNCBD) was designed to be a 'practical tool for translating the principles of Agenda 21 into reality... The Convention recognizes that biological diversity is about more than plants, animals and micro organisms and their ecosystems – it is about people and our need for food security, medicines, fresh air and water, shelter, and a clean and healthy environment in which to live' (UN, undated). Specifically regarding women, the UNCBD preamble included text emphasizing 'the vital role that women play in the conservation and sustainable use of biological diversity and affirming the need for the full participation of women at all levels of policymaking' (UNCBD, 1992), and one of the subsidiary bodies recognized women's knowledge, practices and gender roles in food production (UNCBD, 1996).

Box 7.3 *Women's influence on the Rio Conference outcome*

The Earth Summit in Rio de Janeiro was momentous for the international women's movement. Women activists and advocates, including government representatives and participants from numerous UN agencies, rallied for a strong outcome for gender equality. While the Earth Summit or UNCED lasted from 3–14 June 1992, it was the culmination of years of planning. The process began as early as 1989, with civil society preparing for the conference. Prior to the Earth Summit, the Women's Environment & Development Organization (WEDO) brought together more than 1500 women from 83 countries to work jointly on a strategy for UNCED from 9–12 November 1991 in the World Women's Congress in Miami, Florida (see also Chapter 8). The result of this major event was Women's Action Agenda 21, an outline for a healthy and peaceful planet that was the basis for introducing gender equality into two of the official final documents, Agenda 21 and the Rio Declaration. A demonstration of civil society, government and UN representatives working toward a common goal, the World Women's Congress and the Women's Tent, or *Planeta Femea,* during the Earth Summit in Rio de Janeiro were a milestone for gender and environment policy-making.

Also at UNCED, the Convention to Combat Desertification (UNCCD) was drafted, and then later adopted in 1994. The UN community had begun seriously addressing desertification and issues of land degradation in the 1970s but, according to the UN Environment Programme (UNEP), by 1991 the problem was actually worsening at the global scale. Like the UNCBD, the UNCCD was a comprehensive framework by which the global community could pursue sustainable development goals and included key texts on gender, recognizing the specific roles, impacts, expectations and knowledge of women and men to the issue of desertification. For example, Article 10 of the UNCCD mandated that national action programmes shall 'provide for effective participation at the local, national and regional levels of non-governmental organizations and local populations, both women and men, particularly resource users, including farmers and pastoralists and their representative organizations, in policy planning, decision-making, and implementation and review of national action programmes' (UNCCD, 1994).

While two of three Earth Summit conventions advanced gender perspectives, the third – the United Nations Framework Convention on Climate Change (UNFCCC) – did not include gender aspects. The UNFCCC, addressing one of the most comprehensive issues of this era, was negotiated during the UN General Assembly in 1990 and signed in Rio two years later. The UNFCCC, which entered into force in 1994, serves as the major framework for tackling the causes of and impacts from a changing climate. Its 1997 Kyoto Protocol, with an exclusive focus on mitigation strategies, set binding emission reductions targets for 37 industrialized countries and recognized 'common but differentiated responsibilities' between the developed countries and those countries with developing economies or in transition. The UNFCCC elaborated three market-based mechanisms – emissions trading, the Clean Development Mechanism (CDM) and Joint Implementation (JI) – designed to help developed nations lower their emissions of greenhouse gases (GHGs) (UNFCCC, undated). The first phase of the Kyoto Protocol will expire in 2012, and the parties to the UNFCCC are negotiating a post-2012 framework that extends beyond mitigation strategies to comprehensively address the causes and impacts of climate change, including adaptation, technology transfer, capacity-building and finance. Gender advocates are actively engaged in the negotiation process advocating to mainstream gender into the new agreement.

Box 7.4 *Gender advocates increase status in climate talk forum*

While the UNFCCC instrument neglects to specifically include social dimensions – much less gender specific aspects – of climate change, 2009 represented a major shift in the process. In a turning point for gender and climate change history, the UNFCCC body formally recognized women's civil society groups as a Provisional Constituency to the process. Although the status is provisional, it is presumed the constituency will meet the requirements and become securely established as an official group. Business and environmental organizations had long been a part of climate discussions as official major groups, but because the UNFCCC had not ensured space for civil society in the same way that other UN processes had – such as the Commission on Sustainable Development that recognizes the nine 'major groups' identified by Agenda 21 – women struggled to secure recognition as a major group to the UNFCCC. Women's NGOs rallied together to form a constituency and, along with farmers, youth and indigenous peoples, became one of the more recent formal fora for civil society engagement in the UNFCCC process. The UNFCCC remains a party driven process but civil society input is important and, as a constituency, the group is able to make official interventions on the floor on behalf of women and gender equality (see also Chapter 8).

Resolutions

Among the numerous resolutions adopted by the General Assembly, the Security Council and subsidiary bodies of the UN, a few are integral to emerging policies in the arena of gender and climate change.

Reaffirming the 1997 Resolution E/1997/66 call for widespread gender mainstreaming into every aspect and agency of the UN system, in 2005 the UN Economic and Social Council (ECOSOC) adopted Resolution 2005/31. The resolution indicated that advancing gender equality was fundamentally important to meeting development goals, especially in applying the Beijing Platform for Action (UN ECOSOC, 2005).

In 2000, the UN Security Council issued Resolution 1325 calling for a gender perspective in peacekeeping operations and for the full participation of women in promoting sustainable peace (UN Security Council, 2000). It was reaffirmed and strengthened by Security Council Resolution 1820 (2008) to include a mechanism to report violations and to urge sanctions for violations (UN Security Council, 2008a). While the resolutions do not specifically address or incorporate climate change issues, some security experts have expressed concern that climate change may exacerbate or even cause conflict and thus the issue of women's security in these cases is tangentially related to the gender and climate change agenda (UN Security Council, 2008b).

Most recently, in 2009, the UN General Assembly reaffirmed the sentiment of the 1988 Resolution 43/53: 'Protection of global climate for present and future generations of mankind' (UN General Assembly, 1988). Expressing deep concern about the impacts of a changing climate, and recognizing the momentous December 2009 Copenhagen climate conference, nations' representatives to the General Assembly unanimously adopted the Resolution. The reaffirmation acknowledged women as key actors in sustainable development and recognized that a gender perspective can aid in addressing climate change (UN General Assembly, 2009).

Declarations, plans of action, and other frameworks

Agreements in this category include a broad spectrum in terms of subject matter and leverage. Here the most relevant declarations, plans of action and other frameworks are listed in chronological order to represent how the agreements build upon each other and at times address gaps in previous agreements. The MDGs and the results of the Beijing Conference are considered more influential in global policy than others but all are regularly referred to by those working in the gender and climate change arena.

One of the earliest declarations to inform the gender and climate change agenda is the 1948 UN-UDHR. While the 1945 UN Charter 'reaffirms the equal rights of men and women', including eligibility to participate in UN organs, the UN-UDHR – as the first UN resolution solely dedicated to define human rights – deepens the call via specific articles on the right to own property, freedom of movement and equal protection before the law (UN, 1948). These are rights which are vital for all people and particularly for women as they adapt to the impacts of climate change.

The UN Declaration on the Rights of Indigenous Peoples (DECRIPS), adopted by the General Assembly in September 2007, is a rights-based statement recognizing the rights of indigenous peoples to, among other things, self-determination, economic and political development, land and education, with due respect to customs and systems of indigenous peoples. In particular, Article 21 of DECRIPS reads 'states shall take effective measures... to ensure continuing improvement of [indigenous peoples'] economic and social conditions', with 'particular attention' to the rights of indigenous women, among other demographic populations; and Article 22 urges states to take measures to ensure indigenous women's protection against violence and discrimination (UN General Assembly, 2007). The Declaration's application to the gender and climate change agenda lay in that by ensuring the above rights, women and indigenous peoples increase their resilience to climate change impacts as well as their legal rights to land, which could become increasingly relevant in relation to upcoming mitigation mechanisms such as Reducing Deforestation and Forest Degradation in developing countries (REDD and REDD+, see Chapter 6).

The Beijing Declaration and Platform for Action, the two outcomes of the 1995 Fourth World Conference on Women, are agreements arrived at by governments and the UN to promote a gender perspective in development policies and programs at all levels – local, national and international (UN DPI, 1997). The Declaration links to sustainable development and climate change more broadly by addressing land and credit policies (UN, 1995a). Strategic Objective K of the Platform links more specifically through its aim to address women's issues related to the environment through advancing women's participation in decision-making and including gender perspectives in sustainable development policies and assessments of the impact of development and environment policies (UN, 1995b).

In 2000, countries virtually unanimously adopted the Millennium Declaration, committing to women's empowerment and gender equality as a strategy to achieve sustainable development. The eight related MDGs include the goals of gender equality, environmental sustainability and poverty eradication. The MDGs have been criticized for a top-down approach to the issues and for their weak indicators, however the latter have been reviewed and strengthened since their launch. Additionally, while the

Declaration aims 'to promote gender equality and the empowerment of women as effective ways to combat poverty, hunger and disease and to stimulate development that is truly sustainable' (UN General Assembly, 2000), the goals themselves lack clear overlap or integration, thereby reducing their effectiveness. Nonetheless, they represent an international intention to address the roots of poverty by 2015, and their monitoring and reporting stimulate awareness and action on multiple fronts.

The Hyogo Framework for Action (HFA) resulting from the 2005 World Conference on Disaster Reduction includes clear recognition of links between gender equality and disaster risk, which can be applied to disasters related to climate change. The HFA calls for a gender perspective in disaster risk-management policies, plans and decision-making processes, which includes incorporating gender considerations into early warning systems and providing women equal access to training (ISDR, 2005). At the same time the HFA promotes the integration of disaster risk reduction/management strategies with climate change adaptation strategies, recognizing that climate change can exacerbate disasters and both climate change and disaster can contribute to the vulnerability of peoples.

Conclusion

Summary of gender and climate change instruments to date

The instruments outlined in this chapter are by no means exhaustive. Rather, they reflect the principal agreements to date that are most frequently drawn upon by policy-makers, government leaders, development professionals and advocates to inform emerging policies and practices on gender and climate change. Thus they are shaping international norms and hold potential to inform a promising gender and climate change regime. In some agreements the linkages between gender and climate change are fairly explicit (HFA, for example), while in other cases the linkages must be teased out through application and analysis (for example, UNFCCC). This piecemeal process is a challenge, however several common threads run through all these instruments and can be considered pillars of the emerging gender and climate change framework: (a) Equal rights and access to resources such as land and credit; (b) Participation in decision-making processes; (c) Priority to women, especially the most marginalized, for capacity-building and addressing risks due to exacerbated inequalities; (d) Governance of climate mechanisms that are just and accountable to

Box 7.5 *Commission on the Status of Women (CSW)*

In 1946, one year after the founding of the UN, the UN Economics and Social Council established the Commission on the Status of Women (CSW), responding to demands by many women and gender equality advocates. Since its inception, the CSW has served as a forum for global policy-making on issues related to gender equality. Its rich history has had a profound influence on the development of human rights and gender equality policy and has recently made significant contributions to the global gender and climate change framework. In its decades of work, the CSW's mandate has expanded to encompass promoting women's rights and the objectives of gender equality, development and peace in all spheres; monitoring the implementation of measures related to women's advancement; reviewing progress at all levels (global to national); periodically reviewing the Beijing Platform for Action; informing gender mainstreaming at the UN; identifying emerging issues and trends; and making recommendations on issues needing urgent attention. The Commission has met regularly since 1947, when all 15 representatives were women, and since then has developed a working relationship with NGOs that has led to the incorporation of civil society contributions into CSW decisions. Early CSW members also forged ties with the human rights regime and successfully incorporated more inclusive language into the UN-UDHR. Over the years, the CSW has successfully addressed women's advancement in terms of political, marriage and pay rights, development and security. It was instrumental in the 1979 CEDAW and the 1995 4th World Conference on Women, held in Beijing, which is widely recognized for its participatory nature and strategic approach to advancing women's status in international policy. Most recently, the CSW has taken on the issue of climate change, in 2008 declaring 'Gender perspectives in climate change' its key emerging issue urging governments to integrate a gender perspective into environmental policies and provide adequate resources to ensure women's full and equal participation in decision-making. In 2010, gender and climate change have been incorporated into numerous side events at the annual CSW session, underscoring the importance of the CSW's participation in a gender-responsive global climate policy (CSW, undated).

women and men; and (e) Mainstreaming gender in all levels of climate-related programming design, development, implementation, monitoring and evaluation.

While these instruments lack mechanisms to robustly guarantee implementation, enforce compliance or address impunity, they have normative power to shape the political, economic and development landscape by consciously and publicly placing priorities on paper. The instruments have secured endorsement by a majority of states – in many cases by virtually universal support – which contributes to their legitimacy in the international sphere. Moreover, the content of many instruments drew upon significant input from civil society, which can contribute to up-take at the local or national level in domestic policies and laws, as well as serve as a catalyst for more inclusive policies on the ground.

Significance of gender equality to climate change agenda

Gender inequality is increasingly recognized as one of the last hurdles to sustainable development and the realization of human rights. It is poised to be a defining development objective of this century, alongside addressing climate change. Indeed the two are highly interdependent. Climate change is an urgent issue in and of itself; its impacts and response measures are expected to exacerbate existing inequalities and truly thwart sustainable development. At the same time, ensuring women equal rights and promoting the role of women as capable leaders provides a clear opportunity to positively advance sustainable development and climate policy goals, while not doing so puts their empowerment and lives at higher risk. That these two issues are resonating in the public and political spheres at the same time and with the same amplitude provides an opportunity for them to reinforce and advance the other. As a global normative framework is coalescing to advance these two in tandem, it has tremendous potential to stimulate and reinforce efforts at the national and local level. Already, government ministries, and experts and organizations advocating for women, indigenous and environmental rights, as well as other stakeholders, are drawing upon these instruments and combining efforts to develop locally appropriate climate policy and responses.

Nothing exists, however, in isolation. The future for gender and climate change policy rests in great part on political will at all levels of the international and local community. However, the foundation for a normative framework exists and offers the global community an opportunity to mindfully piece together human rights, a healthy environment and sustainable development to more comprehensively address the challenge of gender inequality and climate change.

References

CEDAW (Committee on the Elimination of Discrimination against Women) (1979) *Convention on the Elimination of All Forms of Discrimination Against Women,* Office of the UN High Commissioner for Human Rights, www2.ohchr.org/english/law/cedaw.htm, accessed 29 January 2010

CEDAW (2009) *Statement of the CEDAW Committee on Gender and Climate Change*, www2.ohchr.org/english/bodies/cedaw/docs/Gender_and_climate_change.pdf, accessed 2 February 2010

CSW (Commission on the Status of Women) (undated) *Short History of the Commission on the Status of Women*, www.un.org/womenwatch/daw/CSW60YRS/CSWbriefhistory.pdf, accessed 12 February 2010

Harris, J. M. (2006) *Environmental and Natural Resource Economics: A Contemporary Approach*, Houghton Mifflin, Boston, MA

ISDR (International Strategy for Disaster Reduction) (2005) *Hyogo Framework for Action 2005-2015: Building the Resilience of Nations and Communities to Disasters*, World Conference on Disaster Reduction, Kobe, Japan, www.unisdr.org/eng/hfa/hfa.htm, accessed 31 January 2010

UN (1948) The Universal Declaration of Human Rights, www.un.org/en/documents/udhr, accessed 12 February 2010

UN (1995a) *Fourth World Conference on Women: Beijing Declaration*, www.un.org/womenwatch/daw/beijing/platform/declar.htm, accessed 30 January 2010

UN (1995b) *Fourth World Conference on Women: Platform for Action*, www.un.org/womenwatch/daw/beijing/platform/index.html, accessed 30 January 2010

UN (2008) *United Nations Treaty Collection; Treaty Reference Guide*, http://untreaty.un.org/English/guide.asp, accessed 29 January 2010

UNCBD (UN Convention on Biological Diversity) (1992) *Multilateral Convention on Biological Diversity (with annexes)*, Rio de Janeiro, Brazil, www.cbd.int/doc/legal/cbd-un-en.pdf, accessed 30 January 2010

UNCBD (1996) *SBSTTA 2 Recommendation II/7 at Second Meeting of the Subsidiary Body on Scientific, Technical and Technological Advice*, Montreal, Canada, www.cbd.int/recommendation/sbstta/?id=6998, accessed 30 January 2010

UNCCD (UN Convention to Combat Desertification) (1994) *United Nations*, www.unccd.int/convention/text/pdf/conv-eng.pdf, accessed 29 January 2010

UNCED (UN Conference on Environment and Development) (1992) *Agenda 21*, UN, www.un.org/esa/dsd/agenda21, accessed 30 January 2010

UN DPI (UN Department of Public Information) (1997) *Fourth World Conference on Women (1995)*, www.un.org/geninfo/bp/women.html, accessed 30 January 2010

UN ECOSOC (UN Economic and Social Council) (2005) *Resolution 2005/31; Mainstreaming a Gender Perspective Into All Policies and Programmes in the United Nations System*, www.unhcr.org/refworld/docid/463b3d652.html, accessed 3 February 2010

UNFCCC (UN Framework Convention on Climate Change) (undated) *Kyoto Protocol*, http://unfccc.int/kyoto_protocol/items/2830.php, accessed 20 January 2010

UNFPA and WEDO (UN Population Fund and Women's Environment & Development Organization) (2009) *Climate Change Connections: Gender, Population and Climate Change*, resource kit, UNFPA and WEDO, New York

UN General Assembly (1988) 'Resolution 43/53; Protection of global climate for present and future generations', www.un.org/Depts/dhl/res/resa43.htm, accessed 20 January 2010

UN General Assembly (2000) 'Resolution 55/2; United Nations Millennium Declaration', www.un.org/millennium/declaration/ares552e.htm, accessed 30 January 2010

UN General Assembly (2007) 'Declaration on the Rights of Indigenous Peoples' www.un.org/esa/socdev/unpfii/en/drip.html, accessed 20 January 2010

UN General Assembly (2009) 'Resolution 63/32; Protection of global climate for present and future generations', www.unece.org/trans/doc/2009/themes/A-RES-63-32e.pdf, accessed 30 January 2010

UN Security Council (2000) 'Resolution 1325', www.un.org/events/res1325e.pdf, accessed 29 January 2010

UN Security Council (2008a) 'Resolution 1820 (2008)', www.un.org/Docs/sc/unsc_resolutions08.htm accessed 1 February 2010

UN Security Council (2008b) 'Security Council demands immediate and complete halt to acts of sexual violence against civilians in conflict zones, unanimously adopting Resolution 1820 (2008)', New York, www.un.org/News/Press/docs/2008/sc9364.doc.htm, accessed 1 February 2010

8

Why More Attention to Gender and Class Can Help Combat Climate Change and Poverty

Gerd Johnsson-Latham[1]

Introduction

The general illustration of climate change is often melting ice and threatened livelihoods of species such as polar bears, and focuses on adaptation to cope with aggravating threats. However, in order to address climate change it is important to be aware of the causes of greenhouse gas (GHG) emissions, and to analyse how human action and power-structures have been decisive for governments – globally – in choosing more and more energy-intensive paths to 'development' and 'well-being'. This chapter looks at how dominant models for development have implied material wealth for many, particularly rich males, while evidence points to the fact that these paths threaten the climate. Furthermore, these models have not addressed root causes to female ill-being such as male violence, lack of control of women's own lives and sexuality, and their exclusion from decision-making. Nor have the paths chosen for development been able to alleviate persistent, widespread poverty, particularly for women, measured both in economic terms, ill-health, insecurity and exclusion. Thus, while acknowledging that unsustainable lifestyles do have many faces in terms of ethnicity, age, disabilities and so on, this contribution focuses on gender and class by addressing excess consumption of the richest – notably men – and the abject poverty of the poorest – notably women.

This chapter builds primarily on previous work, published in 2007 in *A study on gender equality as a prerequisite for sustainable development: What we know about the extent to which women around the world live in a more sustainable way than men, leave a smaller ecological footprint and cause less climate change* (Johnsson-Latham, 2007). The study, Report 2007:2 of the Environment Advisory Council in Sweden, was presented at the UN Commission for Sustainable Development, May 2007, in New York.

Global inequalities and gender-based inequalities

Economic growth is a recent phenomenon in human history, and began with the Industrial Revolution in Britain at the end of the 18th century, as stressed in *Growth Report: Strategies for Sustained Growth and Inclusive Development of the Commission of Growth and Development*, published by the World Bank in 2008. The report however concludes that today, 'growth is a necessary if not sufficient condition for broader development' (World Bank, 2008, p1). Thus, annual reports and forecasts of the International Monetary Fund (IMF), the World Bank, the Organisation for Economic Co-operation and Development (OECD) and the European Union (EU) place economic growth as a centrepiece in development.

Widespread economic growth in recent decades has led to a decline in the number of poor people in the world. In 2005, about 20 per cent of the global population lived on less than US$1 a day (World Bank, 2004). Women remain the poorest of the poor, both in financial terms and in terms of a lack of basic rights such as the right to one's own body, the right of inheritance and the right to property, education, mobility and respect (UN, 2000).

Today, the global population stands at 6.6 billion (IEA, 2009). According to the UN, in 2050 world population is projected to increase to more than 9 billion (UN, 2008). Economic growth has been rising worldwide, with a sevenfold increase recorded since 1950, which is bringing further pressure to bear on the Earth's resources (World Bank, 2006). Furthermore, growth has often been unevenly distributed (UN, 2005).

The differences between countries in terms of energy consumption are staggering. One billion of the Earth's 6 billion inhabitants use 75 per cent of all energy and account for the bulk of all emissions from industry toxins and consumer goods (WWF, 2008). The US and Canada, with 5 per cent of the global population, accounts for 27 per cent of oil consumption. Europe, with around 10 per cent of the population, accounts for 24 per cent. The consumption of fossil fuels in fast-growing economics such as the BRIC-countries (Brazil, Russia, India and China) increases dramatically. The average person in the US uses about twice as much resources as the average person in the EU. Meanwhile, the EU member states together use more resources than the whole of the population of Asia, which is 4–5 times the size of the EU population (IEA, 2005).

The 2006 report on fulfilment of the UN's Millennium Development Goals (MDGs) states in reference to Goal 7, relating to sustainable development, that in 2003 the developed countries (with about one sixth of the global population) accounted for about half of the world's total emissions – 12,106 million tons of CO_2 out of a total of 25,168 million

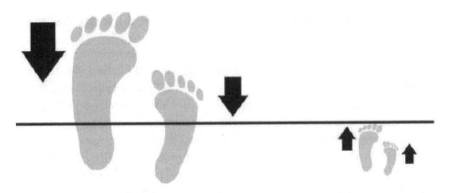

Figure 8.1 *Ecological footprints of men and women in rich and poor countries*

(Left footprint: male; right: female; left side: rich country; right side: developing country)

tons. The report also noted that emissions per capita in the same year were three times as high for rich countries as for poor countries, totalling 12.9 and 4.0 respectively (UN, 2006).

The figure above illustrates the ecological footprints of men and women in rich populations (left) compared to poor men and women (right), as well as differences between men and women.

Average CO_2 emissions in tons per capita in 2004 were as follows: United Arab Emirates, by far the highest with 34, followed by the US with 20.6, Canada with 20, the Ukraine with 16.4, Australia with 13.6, Russia with 10.6, Japan with 9.9, and Germany, the UK and South Africa each with 9.8. The average figure for sub-Saharan Africa was 1.0 (IEA, 2009).

A global 'consumer class', primarily middle class, is to be found worldwide, displaying similar preferences and consumption patterns – and gender-specific differences. Like other groups, this large group of middle-class consumers is constantly influenced by advertising and by popular culture that show what is expected of 'a real man' and 'a real woman' in terms of relationships, lifestyles and consumer choices (Nyberg and Sto, 2004).

For rich countries and groups, the goods and services on offer in a changing, globalized world give individuals an ever-increasing range of options. This is particularly true of men, who enjoy greater freedom than women throughout the world. Consumer choices are seldom 'voluntary', even for the rich, but are shaped by social and other conditions, by economic resources, and by health, age, gender and other factors. Consumption has become a way of expressing group affiliation, status and also represents expressions of 'self-identity' (Giddens, 1991).

A study of the richest may also give keys to understanding global consumption patterns. In 2007, according to Forbes, the world has almost

1000 people with fortunes in excess of US$1 billion (www.forbes.com). Although a number of these billionaires may lead anonymous lives, the luxury consumption of rich people has a profound impact on other people all over the world, as a result of advertising, films and other forms of product promotion. As men are the dominant group among decision-makers in the business community, in politics, in the media and in sport, and are often trendsetting icons, it would be interesting to analyse how men – not least rich men – influence the global direction of consumption. The contents of several 'lifestyle' magazines and commercials give an idea of what direction this is: often they encourage greater consumption of luxury goods such as cars, boats, computers, technology, foreign travel, and so on (*Best Life*, 2007).

Women's shopping is much talked about. It would be interesting to explore what kinds of sums and CO_2 emissions this involves in comparison with the purchases of men which are more likely to involve expensive capital goods such as home electronics, computers and the like, as well a petrol-hungry yachts, motorbikes and motorcars.

The need to provide for the resources required for consumption around the globe makes countries and governments act through a number of means; trade policies, diplomacy and if need be military means, as discussed among others by Klare (2001) in his book *Resource Wars: A New Landscape of Global Conflict*.

The case of Sweden

Due to the fact that the gross domestic product (GDP) has risen by 75 per cent in Sweden since 1975, consumption has increased substantially over this period, and many Swedes have enjoyed access to goods and services that could be placed in the 'bonus consumption' category. Households have thus increased the share of consumption that they use for communications, leisure activities, entertainments and culture, recreation, charter travel and visits to cafés and restaurants. Also, people in Sweden live in increasingly large homes and travel 50 per cent more than they did 25 years ago, primarily to and from work, school and the shops. Over the past ten years, air travel has increased significantly (Ministry of Agriculture, 2006).

In Swedish society, individual opportunities and circumstances are strongly influenced by the person's gender, as consumption is distinctly gender-related. Women, for example, are more likely to purchase basic essentials in the form of less expensive but recurring consumer goods for the whole family, such as food, clothing and household articles, while men are more likely to buy expensive capital goods and own things such as homes, cars and home electronics. Also, men are more likely to eat out than women, and consume more alcohol and tobacco than women (Ministry of Agriculture, 2006).

Distribution is uneven, and men earn and consume more than women in Sweden (Nyman, 2002; Swedish Government Long-Term Planning Commission, 2003). Many of the products found on the market today are created to correspond to the bodies, interests and needs of one sex or the other. Human consumption also reflects the fact that women in general earn less than men and have less money at their disposal (Ministry of Agriculture, 2006).

The case for poor women and men

Worldwide, about half a billion people live in what is termed the 'survival economy' with limited use of the Earth's resources and energy compared with the consumption of the rich. Low consumption and limited CO_2 emissions of the poor could be described as the 'salvation of the rich' in that poor people deplete our common resources to such a limited extent.

In sub-Saharan Africa, only 5 per cent of the population has access to electricity. This often affects women more than men, as the home is usually their workplace (also for women who have paid employment). In the absence of electricity they have neither water pumps nor lighting and they use stoves that cause the death of some 1.5 million women and children every year as a result of respiratory illnesses, a woman's health issue that still remains low on the global health agenda (WHO, 2009). Lack of electricity and of safe, accessible transport also reduces women's chances of obtaining education and training, medical care or a paid job, or of being part of a network. Above all, lack of modern forms of energy means that women's work is heavy and time-consuming, which reduces their chances of moving out of poverty and taking part in decision-making on the same terms as men, who often have more leisure time than women (Johnsson-Latham, 2004).

Gender gaps tend to be greatest among poor families (World Bank, 2001). The family seldom represents the balanced distribution unit on which economists frequently base their approach (Kabeer, 2003). Rather, women and girls eat last and least and are given less money, less education and less space. The World Bank emphasizes that female poverty affects both women in general and society as a whole, since growth is hampered by it (World Bank, 2001).

Mobility for men – on unequal and unsustainable transportation

Transportation of people and goods represents one of the largest and fastest growing sources of GHG emissions such as CO_2, which in turn substantially affect the Earth's climate. Transport has a clear gender perspective as the free movement of individuals is a matter of power and resources. Throughout history, people's capacity for moving around – their mobility – has been crucial to their chances of taking control of their own lives, making a living and satisfying their craving for freedom. People have travelled far and wide – but the travel stories have almost always concerned men: conquerors such as Genghis Khan, explorers such as Christopher Columbus and scientists such as Carl Linnaeus.

Through the ages, women have had less freedom than men – not just mothers, but women in general. In modern times, however, the improvement in women's sexual, reproductive, and economic and social rights has changed the picture and increased their mobility and thus their access to both wage labour and participation in decision-making at various levels. For hundreds of million women however, the only option still is to stay where they are, walk or use the simplest and cheapest methods of transport, such as the bicycle.

Figure 8.2 illustrates how men are much more mobile globally than women, and that their mobility causes more CO_2 emissions and attains far more investments.

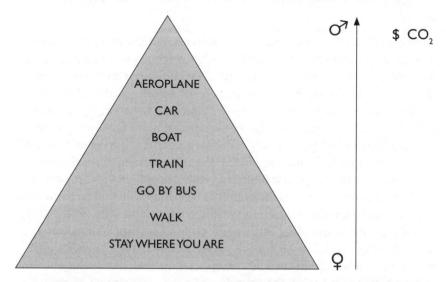

Figure 8.2 *Mobility, CO_2 emissions and investments among women and men*

Mobility is increasingly about 'automobility'

In Sweden, women represent a majority – two thirds – of all households lacking a car/driving licence. About 40 per cent of all drivers (male and female) travel less than 2.5km a day on average. Men drive cars more than women, among other reasons because they commute more widely. A very small group – 10 per cent of all car drivers, primarily men – account for 60 per cent of all car driving in Sweden and thus for the same proportion of emissions and environmental impact. It is estimated that men account for some 70 per cent of all car driving in Sweden, expressed as person kilometres (Swedish National Road Administration, 2005).

The Swedish National Road Administration's report *Res jämt* ('Equal Travel') observes that women are brought up to be sensitive to the needs of others and not to put their own interests first, while men are brought up to show courage and take risks – a situation that is clearly reflected in the way the two sexes behave on the road. Male risk-taking is connected to the fact that men in general regard themselves as independent individuals while women tend to have much larger capacity to consider the needs of a group. Also, men drive the largest and most petrol-hungry vehicles. The report points to the fact that infrastructure-planning at large is undertaken by middle-aged male engineers, by men for men (Swedish National Road Administration, 2005).

Large-scale investment in roads, automobiles and air travel

Men also fly to a greater extent than women and are in a clear majority as regards business travel on regular flights, which consume more energy as they give more space to each traveller compared to charter flights. Women, on the other hand, use public transport – bus and rail travel – to a greater extent; they also travel by air, but then largely on charter trips for holidays. Given the above, women are more favourably inclined towards and more dependent on public transport than men (Swedish National Road Administration, 2005). For most women the car is a means of transport, while for men it is often a symbol on which they are prepared to spend much of their (and their family's) consumption budget in order to demonstrate their status, attitude, wealth, courage and taste (Polk, 2001).

Government spending in Sweden on land transportation stands at 381 billion Swedish kronor for the period 2004–2015. The major part is allocated to roads, railways and transportation that serves the business community, according to the overall goal set by parliament in regard to transportation. Thus, only a limited portion of the budget is allocated to public transportation of persons, including women, children and the elderly, who dominate as users of public transport, given the fact that they have far less access to cars than men (Swedish National Road Administration, 2005).

Creating space for gender equality in climate change policies

In an article in the magazine *Women and Environment*, Minu Hemmati and Ulrike Röhr (2007) ask if there 'Ain't no space for gender in climate change policy?'. Hemmati and Röhr stress that integrating gender analysis can add to the quality, effectiveness, legitimacy and likelihood of implementing climate protection policies and can help meet public transportation needs without using private cars. They also point to the fact that absence of gender analysis in climate related policy-making, climate protection measures may increase the inequities for women. For example, replacing fossil fuels with renewable sources of energy, to maintain current patterns of 'automobility' in richer countries, can create additional burdens on poor women who may need to walk even longer than today to collect fuelwood (World Bank, 2009).

Hemmati and Röhr's observation has become even more relevant as the climate negotiations in Copenhagen in December 2009 focussed considerably on adaptation measures in poor countries to handle climate change. Rich countries have promised to bear some of these costs, by contributions to a global fund for adaptation. However, many rich polluting countries prefer these expenses, compared to both economic and political costs for mitigation measures back home to curb emissions of GHGs in their own countries. Thus, the solution advocated to climate change is primarily adaptation, among the poor in the southern hemisphere, rather than shifts in high-carbon lifestyles in the Western world.

Over the years, both researchers and women's organizations and groups have stressed gender equality in relation to sustainable development, energy, environment and more recently climate change. Gender has been stressed both as part of sustainability and as key to addressing unsustainable paths of development. At the global level, the UN Commission on Sustainable Development (CSD) and the United Nations Environment Programme (UNEP) have created space for groups/networks such as ENERGIA and the Women's Environment & Development Organization (WEDO). These organizations have emphasized the importance of tackling overarching issues involving male bias – by boosting women's participation in decision-making forums, women's right to own and inherit land, education and capacity-building. They have also been persistent in pointing at technologies that could protect women from respiratory diseases, a major cause of death in poor countries (WHO, 2009).

Key contributions to sustainable ways to development and human well-being have also come from the broad gender equality agenda. A major case in point is – still – the action plan adopted at the 1995 UN Fourth World Conference on Women in Beijing. The Platform for Action provides a wide, rights-based agenda in which economic well-being is only one of a number of dimensions, alongside basic issues (also raised in Agenda 21, Rio 1992) such as health, sexual and reproductive health and rights (SRHR), action against violence, education, participation and voice, the rights of the girl child, women's right to land. The gender equality agenda thus embraces all three dimensions of sustainability as defined by the UN: ecology, economy and social sustainability. Thus, by applying a gender equality perspective, a welfare model may be developed that is addressing both women's needs and paths to sustainable development (UN, 1995).

Conclusions

This chapter has aimed at bringing out often-neglected facts and pointers concerning dissimilarities in the lifestyles and consumption patterns of women and men, and thus in their environmental impact. It has done so by pointing at how men (notably rich men), through their greater mobility and more extensive travel, account for more CO_2 emissions than women, in both rich and poor countries. The chapter points at how women and female lifestyles can be part of the solution to climate change, instead of women being considered as victims of inevitable developments and problems that will have to be dealt with – later, once the overall framework for work on climate change has been set. Such a perspective reverses the gaze by stressing that a key aspect of gender and climate change is a changed behaviour particularly among (rich) men, thereby enhancing opportunities of the majority of human beings to enjoy sustainable development, both now and in the future.

Note

1   The author alone is responsible for the analyses, proposals and views contained in the text.

References

Best Life (2007) 'What matters to men', theme-issue, November 2007, New York

Giddens, A. (1991) *Modernity and Self-Identity: Self and Society in Late Modern Age,* Polity Press/Blackwell Publishing, Cambridge/Oxford

Hemmati, M. and Röhr, U. (2007) 'Ain't no space for gender in climate change policy?', *Women and Environment,* pp74–75

IEA (International Energy Agency) (2005) *World Energy Outlook 2005,* IEA, Paris

IEA (2009) *World Energy Outlook 2009,* IEA, Paris

Johnsson-Latham, G. (2004) *Power and Privileges,* Ministry for Foreign Affairs, Stockholm

Johnsson-Latham, G. (2007) *A study on gender equality as a prerequisite to sustainable development: What we know about the extent to which women around the world live in a more sustainable way than men, leave a smaller ecological footprint and cause less climate change,* Report 2007:2 of the Environment Advisory Council (Sweden), Stockholm

Kabeer, N. (2003) *Gender Mainstreaming in Poverty Eradication and the Millennium Development Goals: A Handbook for Policy-Makers and Other Stakeholders,* Commonwealth Secretariat, London

Klare, M. T. (2001) *Resource Wars: A New Landscape of Global Conflict,* Henry Holt & Co, New York

Ministry of Agriculture (2006) *Rethink – An Action Plan for Sustainable Consumption,* Regeringsskrivelse, p107, Ministry of Agriculture, Stockholm Ministry of Finance

Nyberg, A. and Sto, E. (2004) *Is the Future ours?,* Statens Institutt for Forbruksforskning, Oslo

Nyman, C. (2002) *Mine, Yours or Ours? Sharing in Swedish Couples,* doctoral thesis, Department of Sociology, Umeå Univiversity

Polk, M. (2001) 'Gender Equality and Sustainable Development: The Need for Debate in Transportation Policy in Sweden', *Transportpolitik i Fokus,* no 1

Swedish Government Long-Term Planning Commission (2003) *Långtidsutredningen,* Ministry of Finance, Stockholm

Swedish National Road Administration (2005) 'Res jämt: tankar kring ett jämställt transportsystem', publikationsrapp, no 110

UN (1992) *Agenda 21, The UN Programme of Action from Rio,* UN, New York

UN (1995) 'Beijing Platform for Action', The UN Fourth World Conference on Women, September 1995, www1.umn.edu/humanrts/instree/e5dplw.htm

UN (2000) *The World's Women 2000,* UN Statistics Division, New York

UN (2005) *World Social Report,* UN, New York

UN (2006) *The Millennium Development Goals Report 2006,* statistical annex, UN, New York

UN (2008) *World Population Prospects: the 2008 Revision Population Database,* UN Population Division, New York

WHO (World Health Organization) (2009) *Women and Health,* WHO, Geneva

World Bank (2001) *Engendering Development: A World Bank Policy Research report,* World Bank, Washington, DC

World Bank (2004) *Measuring Development,* 36th edition, World Bank, Washington, DC

World Bank (2006) *Gender Equality as Smart Economics,* SecM2006-0370, World Bank, Washington, DC

World Bank (2008) *Growth Report: Strategies for Sustained Growth and Inclusive Development*, World Bank, Washington, DC

World Bank (2009) *World Development Report 2010, Development and Climate Change*, World Bank, Washington, DC

WWF (2008) *Living Planet Report 2008*, WWF, Gland

Websites

Ecological Footprint, www.myfootprint.org
ENERGIA, www.energia.org
Forbes, www.forbes.com
IEA, www.international.energy.agency.org
Swedish Government, www.sweden.gov.se
UN, www.un.org
WEDO, www.wedo.org
World Bank, www.worldbank.org
World Business Council for Sustainable Development, www.wbcsd.org

9

Women Organizing for a
Healthy Climate

Irene Dankelman

Accepting the Nobel Peace Prize in Oslo on 10 December 2004, Professor Dr Wangari Maathai mentioned: 'Although this prize comes to me, it acknowledges the work of countless individuals and groups across the globe. They work quietly and often without recognition to protect the environment, promote democracy, defend human rights and ensure equality between women and men. By so doing, they plant seeds of peace.'

This chapter looks at the way women around the world have countervailed environmental crises throughout history. It will pay particular attention to the more recent history of the involvement of women and women's organizations in climate change policy development, awareness-raising and action. Some specific examples of women's activism in the area of climate change are included. Based on such experiences from the past and present, conclusions are drawn on the main drivers of such organizing and action, and on the position of groups of women in social networking for climate justice.

Women's activism worldwide

Already for many centuries women worldwide have manifested themselves as agents of change for a healthy and peaceful planet. As has been mentioned in the introductory chapter of this book, from the history of women's involvement in this area we can learn many lessons. The first documented case of a woman who literally gave her life for safeguarding the environment goes back to 1730 in India. Maharaja Abhay Singh of Jodphur, Rajasthan, wanted to build a new palace and required wood for it. After inviting local men to a party, he ordered his soldiers to the area around the village of Khejadli to fell the khejri trees (*Prosopis cineraria*). When Amrita Devi, member of the Bishnoi community, noticed this, she rushed out and organized the villagers to prevent the cutting down of the trees. She hugged the first tree and said: 'Sar santey rookh rahe to bhi sasto jaan' ('If a tree is saved even at the cost of one's head, it's worth it'). The axe fell on her and she died on the spot. In the massacre, 362 villagers lost their lives while protecting the trees. When the king heard about this, he apologized and promised the villagers that they would never again be asked to provide timber. With that event the recorded history of the Chipko movement in India started (Bishnoi Organisation, 2010).

The Chipko movement in the foothills of the Himalaya (north India) is particularly known for its actions to resist the destruction of community land and livelihoods. In 1974, the government of the state of Uttar Pradesh diverted the men of Reni village (presently in the state of Uttarakhand) to a fictional compensation payment site. At the same time, labourers disembarked from trucks to start logging activities near the village. Under the leadership of Gaura Devi, a 50-year-old illiterate woman, women rushed from their homes to hug the trees ('chipko' means hugging) and prevented the forests from being cut. A four-day stand-off ended in victory for the villagers: the contractors withdrew and the forest was saved. This incident grew into a major grassroots movement of hill women and men, demonstrating their power as nonviolent activists – inspired by Gandhi's philosophy of peaceful resistance. The actions of the women from Reni were repeated in several other places in the country, and the struggles of the Chipko movement continue in many ways – not only preventing further deforestation, but also in preserving the agro-biodiversity on which the agricultural systems depend and ensuring food sovereignty, with more and more emphasis on the changing conditions due to climate change (see Case study 3.1) (Dankelman and Davidson, 1988; UNEP, 2004, 2005).

In Japan, while industrial development had made the society richer, in the 1950s environmental problems began to threaten the health of local citizens. Women started to raise their voices to alert to these and organized an increasingly powerful movement. The Nakabaru Women's Society and Sanroku Women's Society protested loudly against pollution from industries and power plants in the Tobata region. The women held meetings and discussed how to prevent the environmental problems. They also conducted field studies, collecting scientific data on pollution through several years of action research. The women claimed the right to live in a safe and healthy environment. Finally this resulted in significant pollution prevention measures taken by the local government and corporations (Kitakyushy Forum on Asian Women, 1993).

Communities in Africa are facing many environmental challenges, and also in this continent women have played a major role in environmental action. A well-known example of women's long-lasting involvement in environment is the Green Belt Movement (GBM) that originated in

Kenya. At the initiative of Wangari Maathai the GBM was launched by the National Council of Women on Earth Day 1977. This environmental campaign resulted in the mobilization of thousands of women and many men, reclaiming the eroded land by planting indigenous trees. The movement created a national network of 6000 village nurseries, designed to combat creeping desertification, restore soil health and protect water catchment areas. The 50,000 members of the movement have planted about 20 million trees. The GBM has always tried to address issues of women's empowerment in combination with environmental protection, livelihood security and democratization (Maathai, 2006).

'Implicit in the act of planting trees is a civic education, a strategy to empower people and to give them a sense of taking their destiny into their hand, removing their fear...', says Wangari Maathai, founder and leader of the GBM (Maathai, 2003).

In Nigeria, since 1984, women have led remarkable initiatives to stop the exploitation of oil, the massive and dangerous burning of natural gas and the increasing violence in the Niger Delta. Dozens of women's groups across the Delta, mobilized in the multi-ethnic umbrella organization, the Niger Delta Women for Justice, play a major role in keeping big oil companies accountable (Brownhill and Turner, 2006) (see Case study 9.3).

Box 9.1 *Women in Zambia organize for access to water*

Bweenga is a place in the southern province of Zambia, about 30km west of Monze town. For quite a number of years the Southern Province is hit by droughts and unpredictable weather conditions, and the water levels in the wells have gone down so much that most wells have been abandoned. Here more than 45 women, led by 79-year-old Samaria Haanungu, gathered one morning in July 2006, to solve the water problem once and for all. Their main concern was the lack of water for household use. The only place where they could access water was at a dam, where they shared the vital commodity with animals. This meant that women and girls had to get up at 2am to fetch water, and girls often missed school. As they were determined to solve the water problem once and for all, they walked 30km to the council in the hope that they would listen to their concerns and provide them with an answer for their daily problems. The decision to visit the town was fruitful: the council linked the women with the international organization Water Aid, who sank boreholes in their area and also helped them construct ventilated improved pit latrines.

Source: Macha (2010)

In the US, in the aftermath of Hurricanes Katrina and Rita, organizations in the Gulf Coast such as Coastal Women for Change (CWC) and the Women's Health and Justice Initiative (WHJI) have been instrumental in mobilizing and empowering women in recovery and reconstruction. Also women in the matriarchal Native American tribe of the United Houma Nation (UHN) were seriously impacted by the hurricanes. As a woman leader, Brenda Dardar-Robichaux has developed a plan to ease the burden on the families in the UHN (see Case study 5.7).

In the Bolivian Altiplano, the farmers' organization Unión de Asociaciones Productivas del Altiplano (UNAPA) has 60 *yapuchiris*, farmers with a particular artisanal dedication to agriculture. Ten of these are women, and these women *yapuchiris* were storing a very large quantity of potato varieties, grain seeds and other species, including medicinal plants. Knowing under which conditions and where to sow these species and varieties, the women transferred their knowledge to other women farmers. Furthermore, female *yapuchiris* have taken a leading role in negotiating long-term market access for local produce. With changing climatic conditions these *yapuchiris* play an important role in agricultural risk management (ISDR, 2008).

In Eastern Europe it is primarily the pollution of the environment and the impacts on human health and their livelihoods, that have fuelled women's environmental activism. Such movements have been important drivers of democratization processes in the region. Milya Kabirova in 2002: 'Chernobyl stands at the cradle of the Urals ecological movement. My own work is inspired by that of my mother, Sarvar Shagiakhmetova. In 1995 she was the first person to start a lawsuit in order to get recognition of her and our family's diseases linked to radiation and to get compensation from the Mayak nuclear plant. The lawsuit that could have created a precedent for other cases was stopped when my mother died in October 1998 of cancer caused by the accident in Mayak.' In 1999 Milya Kabirova founded the non-governmental organization (NGO) Aigul, which means 'Moon Flower' in Tatar. It is a beautiful name for a sad flower that doesn't grow under the sunshine, but only in the white stillness of the moon, resembling the nuclear winter. The organization's main objectives are to protect the civil rights of people who have been exposed to radiation and of their descendents, to promote an ecological way of thinking, to eliminate nuclear arms production and use, and to promote public participation in shaping state policy and laws (WECF, 2002, pp97–100).

Similar efforts are reflected by the women active in organizations such as MAMA-86, a consortium of 17 women's groups in the major cities of the Ukraine, all working for environmental health and women's empowerment. MAMA-86 was established by young mothers who were confronted by many health problems in the aftermath of the Chernobyl disaster of 1986. Presently main areas of work are promotion of safe drinking water, sanitation and chemical security, lowering the risks for public health and the environment, and the promotion of more ecological lifestyles and products (www.mama-86.org.ua).

International women's organizations and networks

As agents of change, bound together by our commitment to justice, equality and peace, we can sustain our environment and our common future. (WAVE Assembly – Women as the Voice for the Environment, United Nations Environment Programme (UNEP), Nairobi, 2004)

Local, national, regional and international women's environment organizations have mainly been established since the 1980s. Such organizations became important catalysts in empowering women, developing their own analyses and in bringing a gender perspective to the policy level and development practice. These are important fora through which women's voices are speaking out against environmental injustice. Many share the vision that proper gender-mainstreaming is not only a question of women's involvement and participation in shaping sustainable development and environmental decision-making, but also about the contents of development itself. Like former congresswoman Bella Abzug (1920–1998) once said: 'We women do not want to be mainstreamed into a polluted stream. We want the stream to be clean and healthy.'

In the context of climate change the following international women's organizations are playing an important role: the Women's Environment & Development Organization (WEDO), GenderCC, ENERGIA, Women in Europe for a Common Future (WECF), Women Organizing for Change in Agriculture and Natural Resources Management (WOCAN), the Global Gender and Climate Alliance (GGCA) and the Network of Women Ministers and Leaders for the Environment. In this area, the gender programme of the International Union for Conservation of Nature (IUCN) has also been a principal catalyst. Main goals and activities of these organizations are described below. Apart from these organizations also programmes, funds and specialized agencies of the UN have contributed significantly in putting gender and climate change on the international agenda, including the United Nations Development Programme (UNDP), the United Nations Development Fund for Women (UNIFEM), the United Nations Environment Programme (UNEP), the United Nations

Box 9.2 *Women environment leaders*

The initiatives mentioned in this book show that individual women have played crucial roles in mobilizing others and forcing change for a more sustainable future and a healthy climate (Eisler and Corral, 2007). The list of names of these individual leaders at local, national and international level is inexhaustible and this book reflects the leadership of some of them. Just to mention two leaders who unfortunately passed away recently: Joke Waller-Hunter and Khamaruga Banda. As Executive Secretary of the United Nations Framework Convention on Climate Change (UNFCCC) Secretariat **Joke Waller-Hunter** (1946–2005) was one of the most influential people working on climate change at global level and she contributed directly to the development of the Kyoto Protocol. Before joining the UNFCCC, Waller-Hunter was the Director of the Environment Directorate of the Organisation for Economic Co-Operation and Development (OECD) in Paris and she was the first Director of the Secretariat to the UN Commission on Sustainable Development. Before that, she was actively involved in the preparations of Earth Summit (United Nations Conference on Environment and Development (UNCED), Rio de Janeiro, 1992) as Deputy Director at the Netherlands Ministry of Housing, Spatial Planning and the Environment. Waller-Hunter believed in the importance of a strong civil society; this was reflected in her co-founding of the Environment and Development Service Both ENDS in 1986, and the fact that she donated her assets to strengthen leadership in environment and sustainable development organizations in developing countries: the Joke Waller-Hunter Initiative. Aware of the importance of female leadership in environment, she was an active participant in the Network of Women Ministers and Leaders for the Environment (NWMLE) (Both ENDS, 2010).

Another leader, **Khamarunga Banda** (1963–2009) from Zambia is remembered for her tireless efforts to improve the lives of poor women through increased access to energy and addressing energy poverty in Southern Africa. Within the ENERGIA network on Gender and Sustainable Energy (see below) and HEDON Household Energy Network, she was a driving force for exploring new areas of the linkages between gender and climate change, and gender and biofuels. She was equally at home lobbying on international platforms, such as the UNFCCC, as she was doing field-work in a village, and therefore she personally bridged the distance between local and global levels and made rural women's voices heard at international fora. Her excellent writing skills and knowledge of the plight of women form an important source for future action on gender, energy and climate change (www.energia.org and www.hedon.info/1624/news.htm).

For information on some other women leaders on gender and climate change see *Powering Change* from the International Union for Conservation of Nature (IUCN), 8 March 2010, www.iucn.org/?4876?Powering-change--IUCN-celebrates-International-Womens-Day.

Population Fund (UNFPA), the Food and Agriculture Organization of the United Nations (FAO) and the United Nations Educational, Cultural and Scientific Organization (UNESCO).

The **Women's Environment & Development Organization** (WEDO) was established in 1991 by a group of women leaders in environment, including: congresswoman Bella Abzug (1920–1998) and feminist activist and journalist Mim Kelber (1922–2004) both from the US; ecologist and women's rights advocate Wangari Maathai from Kenya; physicist and activist Vandana Shiva from India; activist, journalist and grassroots mobilizer Thais Corral from Brazil; and the activist Brownie Ledbetter from the US. Since its inception, WEDO has been a leader in organizing women for international conferences and activities, following global developments closely, and trying to influence these through its advocacy work. In 1991, WEDO organized the World Women's Congress for a Healthy Planet, bringing together more than 1500 women from 83 countries to work jointly on a strategy for the United Nations Conference on Environment and Development (UNCED). The result was the *Women's Action Agenda 21,* an outline for a healthy and peaceful planet, that formed the basis for introducing gender equality in the official UNCED outcomes, including Agenda 21 and the Rio Declaration. Throughout the 1990s, WEDO played a key leadership role to ensure that gender was included in the major UN conferences. The organization also recognized that policy commitments alone are not enough to improve women's daily lives. That is why WEDO strengthened its collaboration with southern partners on implementing global policy gains at the national level and holding governments accountable to their commitments on women's rights. In the 2000s, the organization advocated strongly for a robust UN gender architecture through the Gender Equality Architecture Reform (GEAR) campaign. In the campaign, co-facilitated with the Center for Women's Global Leadership (CWGL) more than 310 civil society organizations worldwide participated. After working on a wide variety of environment and sustainable development issues, the organization started to focus more and more on the areas of gender justice and climate change. In this field, research is executed, information is developed and distributed, and much emphasis is placed on advocacy for gender mainstreaming in climate change negotiations and actions (see also Chapter 7).

GenderCC – Women for Climate Justice, formally registered as an NGO since 2008, is an organization and global network of women and gender activists and experts from all regions in the world, working for gender and climate justice. GenderCC believes that, in order to achieve women's rights, gender justice and climate justice, fundamental changes are necessary to overcome existing systems of power, politics and economics.

Its goal is to integrate gender justice to climate change policy at local, national and international levels, by raising awareness and building capacity, and increasing the global knowledge base on these issues. Empowerment and increased participation of women are also central to its mission. The secretariat of GenderCC organizes exchanges and global learning, enables publications and supports communication and promotion of gender issues, and maintains an active list-serve and website. One slogan is 'Emissions down, women's rights up', another is 'There is no climate justice without gender justice!'. GenderCC's work is based on the work of regional focal points in Asia, Africa, the Pacific, Latin America, North America and Europe, and the organization collaborates with existing women and gender networks, for example in the Gender Constituency of the UNFCCC, for which it serves as the focal point. Underlining the importance of diversity in GenderCC's work, Titi Soentoro from Indonesia underlines: 'We are a forest. We do not want to be a monoculture tree plantation. We are trees, we are flowers, and our common goal is climate justice.'

ENERGIA is an international organization working on gender and sustainable energy that was founded in 1996. It has grown into an international network of like-minded organizations and individuals, working from the contention that projects, programmes and policies that explicitly address gender and energy issues will result in better outcomes, in terms of sustainability of energy services as well as the human development opportunities available to women and men. ENERGIA's goal is to contribute to the empowerment of rural and urban women through a specific focus on energy. Its regional networks in Africa and Asia are particularly active and vibrant. ENERGIA focuses its efforts, skills and resources around the following strategies: capacity building (especially of project managers and policy-makers); gender in energy projects, showing positive impacts of energy access on livelihoods of poor women and men, and exemplifying how such impacts can be multiplied; policy influencing, working towards gender mainstreaming at institutional and policy levels; and networking, for example with organizations such as WEDO, WECF and the GGCA, of which ENERGIA is a member. The organization has supported the development of gender audits of national energy policies in Nigeria, India, Senegal, Kenya and Botswana (see also Box 10.1). Apart from the promotion of sustainable energy sources ENERGIA pays more and more attention to climate change and gender issues.

Women in Europe for a Common Future (WECF) was created in 1994 following the Rio Earth Summit of 1992 to give women a stronger voice in sustainable development and environmental policy. Presently WECF is a network of more than 100 women's and environmental organizations in 40 countries, spanning Western Europe and the EECCA

(Eastern Europe, the Caucasus and Central Asia) region. WECF has offices in Germany, France and the Netherlands, and its projects in the EECCA region involve some 30,000 beneficiaries. WECF works on safe chemicals and food production, safe water and sanitation, and safe energy and a healthy climate for all. To reach these goals, WECF promotes poverty reduction, public participation and environmental rights, and gender equality. The organization carries out national, regional and international advocacy work on these issues together with its members. WECF works to improve access to affordable and renewable energy in particular for low-income communities by helping partners to develop locally adopted renewable energy, energy efficiency and energy saving technologies, and to access funding mechanisms for large scale replication of best practices for household and community projects. WECF and its partners advocate for safe and sustainable energy and climate justice for all, developing recommendations to improve legislation and practices. It researches the impacts of nuclear energy on human rights and environmental health and safety, and campaigns to redirect nuclear funding to the development and promotion of renewables.

> *My organization works for sustainable rural development in Armenia. Many people in our rural areas suffer from energy poverty... In daily life this often means they can only afford to light or heat one room. To survive, people burn whatever they can: plastic, kerosene etc. without considering health risks. In our project village Hayanist, school children suffer from coughs and other respiratory diseases like asthma because of smoky stoves in the classroom. For me, this is unacceptable! We started to carry out a solar energy project to improve the situation. I believe everyone, including the poor, has the right to sustainable energy.* (Elena Manvelyan, Armenian Women for Health and a Healthy Environment in WECF, 2009, p2)

WOCAN (Women Organizing for Change in Agriculture and Natural Resources Management) is a global network of women and men professionals and farmers in 83 countries who are committed to increasing rural women's access to and control over resources to sustainably manage agriculture and natural resources. WOCAN is part of the advocacy team of the GGCA at global and regional UNFCCC sessions, providing leadership on Reducing Emissions from Deforestation and Forest Degradation (REDD). Current discussions on REDD are very weak with respect to the gender dimensions and its impacts on rural women who have few or no options to use the forest for sources of fuelwood, livestock feed, medicines, and food in times of scarcity. WOCAN advocates for addressing this gap so that the REDD policies, financing mechanisms and consultative processes take full account of the differentiated rights, roles and responsibilities of women and men, promote gender equality and equity, and reward women who protect and manage forest resources (see Chapter 6). WOCAN has launched Payment for Environmental Services pilot projects with rural women's groups in Nepal and Northeast India to tap into carbon markets.

The IUCN Political Declaration of 1998 underlined: 'Gender equity and equality are fundamentals to human rights and social justice fulfillments, and a condition to sustainable development.' After a period of ad hoc initiatives in the area of gender and environment, in 2000 the World Conservation Congress of the **International Union for Conservation of Nature** (IUCN) in Aman approved a resolution that called on its Director General to ensure that 'gender equity is mainstreamed in all the Secretariat's actions, projects and initiatives'. A series of concrete actions, catalysed by the office of IUCN's Senior Gender Advisor, Lorena Aguilar, has accelerated the pace of change. Instruments, tools, mechanisms and advocacy for gender sensitivity in natural resources conservation and sustainable development were promoted by developing guidelines, tools, handbooks and other publications, elaborating gender assessments of organizations and policies, and by developing and delivering training sessions for professionals and policy-makers on gender issues. In major meetings of the United Nations Convention on Biological Diversity (UNCBD), Commission on Sustainable Development (CSD) and UNFCCC it has played a catalytic role in bringing gender issues forward and advocating for women's participation and involvement. Since 2005, IUCN's gender programme specifically focuses its work on gender and climate change and has developed a training model and manual on gender and climate change in the context of the Global Gender and Climate Alliance (see below).

The **Network of Women Ministers and Leaders for the Environment** (NWMLE) was established in March 2002 in Helsinki, Finland. The ministers and women leaders felt an urgency to reverse dangerous trends in the world's development and to address the critical need for visionary and concrete policies towards sustainable development in their own countries and worldwide. Its main objective is to raise awareness of gender aspects within environmental issues globally. Activities include: the development of recommendations for gender mainstreaming in environmental policies and promoting of practical solutions to environmental problems, the creation of critical mass of leadership to influence national and international environmental policies, the exchange of best practices and experiences, and building partnerships with civil society, non-governmental and intergovernmental agencies. Since the Conference of Parties (COP-9) of the UNFCCC in Milan in 2003, during the COPs the NWMLE and its members have been important motors for strengthening of gender language and action. In addition, at the 2008 UNEP Governing Council/Global Ministerial Forum, the NWMLE hosted a High Level forum on Gender and Environment that provided recommendations for the integration of gender perspectives into international environmental governance; green economy and management of chemicals and hazardous wastes. In February 2010, in Bali, Indonesia, the NWMLE launched national and regional chapters on women and environment in order to strengthen the activities of the network at national and regional levels. The current Chair of the NWMLE is Rejoice Thizwilondi Mabudafhasi, Deputy Minister, Department of Water and Environmental Affairs of South Africa. A co-Chair from a country in the north is yet to be nominated. The Secretariat for the NWMLE is situated at the UNEP Headquarters in Nairobi, Kenya.

The **Global Gender and Climate Alliance** (GGCA) was established during UNFCCC's COP-14 in Bali, December 2007. At its launch Julia Marton-Lefèvre, IUCN's Director General, mentioned that efforts to address climate change and gender equality call for collaboration between sectors and institutions. Founding organizations were IUCN, WEDO, UNDP and UNEP, and presently the alliance has grown to include 25 institutions among UN and civil society organizations. GGCA's vision is that incorporation of a gender perspective in all climate change policies is critical to solving the climate crisis. Through advocacy work, information sharing and capacity building, GGCA works to ensure that climate change policies, decision-making, and initiatives at the global, regional and national levels are gender responsive.

What moves women?

The growth of women's power and sustainability of development are ecologically tied. (Workshops on 'Women, Environment and Development', Environment Liaison Centre, Nairobi, 1985)

In the context of this chapter relevant questions are: Why have women become agents of change in the environmental arena in general and on climate change in particular? What specific contributions do they make? And how are they organized?

Although women are not one social category and a slew of differences most be recognized among women themselves – including social class, caste, race, age, educational background and physical locality – what many women worldwide have in common is that they face discriminatory obstacles that hinder their full involvement in and benefit from climate change policies and actions. Main obstacles for gender equality are cultural perceptions and practices, traditional roles and responsibilities (including division of labour), lack of legal frameworks and formal and informal rights and regulations, education that traditionally favours male expertise in climate change science, and institutional cultures and governance structures in the climate change sector that are often male-dominated and in which particularly grassroots women's interests are ill-represented. One of the reasons for the denial of women's participation and rights in the environmental arena is probably that apart from a merely technical and statistical angle, gender carries a strong element of politics and power. As Pietilä (2002) concluded: gender equality requires 'transformative change'. Such fundamental changes often evoke a lot of resistance, at local, but also at national and international levels.

If we try to identify main external drivers and triggers of women's environmental activism over time, similar elements can be observed: threats to livelihoods and livelihood options, dangers to human and environmental health, lack of policy focus on future well-being and that of future generations, the feeling of exclusion from decisions, concern of ineffectiveness and lack of participation. Or – on the positive side – activism in the environmental field is activated by the fact that many women prioritize family and children's health, community and future generations' well-being, prefer to avoid risks, are key caregivers, are sensitive to injustices and often are directly dependent on access to resources of good quality.

> *Women's leadership has an enormous impact in the quest for gender justice. Because women are more open to exchange ideas and reach consensus, they produce horizontal political structures based upon group accomplishment rather than the traditional, hierarchical management structures seen in most parties and unions. Women know how to manage better because of their inherent social abilities: compassion, generosity, understanding and the ability to care for others – we are usually the grand managers of need.* (Sueli Carneiro, Brazil, in Batliwali and Rao, 2002, p15)

Day-to-day work in environmental use and management makes many women an invaluable source of knowledge and expertise. These human capacities create a strong basis for environmental awareness and action. Women's reproductive and productive tasks and responsibilities generate a powerful commitment towards the present and future well-being of children, other family members and communities. This awareness also strengthens their commitment to resist developments or power structures that threaten the subjects of that feeling. Environmental degradation and pollution and inappropriate actions affecting family health and security, increasing women's work burdens, and limiting access to and control over good quality resources, options for well-being and safe livelihoods, trigger mobilization, organization and action by women as individual leaders, groups or movements.

Characteristics

Other important characteristics of most of the individuals, women's organizations and networks mentioned in this chapter, are that they challenge existing policies and practices, striving for a positive social and political agenda, bringing feminist agenda's back to global and local politics. Batliwala (2008, pp12–13) attributes the following qualities to feminist movements as ideal prototype: (a) a gendered analysis of the problem or situation; (b) women form a critical mass; (c) open espousal of feminist values and ideology; (d) women's leadership in the movement reflected at all levels; (e) political goals are gendered – in the sense that they seek a change that privileges women's interests; (f) the use of gendered strategies and methods – building on women's own mobilizing and negotiating capacities; and (g) the creation of a more gendered organization – with a flatter and more collective leadership. Although not all of the organizations and networks mentioned earlier would call themselves 'feminist', most of these characteristics are reflected in their visions, approaches and structures.

Organizations and networks working on gender and climate change at international levels did start off by using strategic spaces and political moments (specifically UN conferences), while those working at local level often come forward out of existing women's organizations and/or from actual or potential livelihood threats. Many share a similar organizational life cycle in which succeeding stages, such as imagining and inspiring, founding and framing, grounding and growing, struggling and learning, and review and renewal can be identified (Batliwala, 2008). A recent development, however, is that of coalition building, in which several NGOs, networks, civil society groups and even UN bodies, come together to strengthen their voice and to have a larger impact on climate change policies and practices. The GGCA is a typical example of that, whereas the women's/gender constituency under the UNFCCC is a particular coalition in which several organizations (GenderCC, WEDO, WECF, LIFE and ENERGIA) share the responsibility for the facilitation of a combined input on gender issues in UNFCCC processes. The question stays, however, if we already can speak about a real social movement on gender and climate change, as these bodies have been established quite recently and are still developing. The relationship between organizations and movements is complex and can even be problematic, meaning it takes time, openness and cooperation to build coalitions with such mass participation that we can call them a movement.

Strategies

The multifaceted strategies applied by the mentioned organizations and networks manifest themselves in different ways: collection of data and information sharing, mobilizing and organizing, developing and refining political analysis and agenda, resistance against negative developments, research and action as watchdogs, advocating for improvement and revision of policies and programmes, political participation, training and capacity building, including educating officials and government representatives, the development of alternative and more sustainable practices, mobilization of resources and strengthening their own organizations and structures, and building relationships and alliances (Arts et al, 2002; Batliwala, 2008). In most of the cases described above, there is a strong emphasis on building 'new' (not exclusively young) leadership. Women are using all these strategies at all levels of society, often aiming for rights-based approaches.

Conclusions

It is noteworthy to mention that, in general, in organizations and individuals working on gender, environment and climate change, one often sees a holistic vision – linking issues such as health and ecology, peace and environment, and human rights and sustainable livelihoods, and in that sense reframing development paradigms. Founder of the environmental organization Wahana Linkungan Hidup Indonesia (WALHI), former Minister of Human Settlement and Regional Development and UN Special Envoy for Millennium Development Goals (MDGs) in Asia and the Pacific, Erna Witoelar, expressed this as follows: 'The ability to see things in a holistic way is an important quality of leadership' (in Batliwala and Rao, 2002, p20). Thereby many frontline organizations and leaders apply boundary work – bridging different spheres and sectors of life (Hoppe, 2010).

The strategies used to reach the actual objective of challenging existing climate change agendas, policies and practices, including mitigation and adaptation, has resulted in: organization of (directly or indirectly) affected women to challenge, resist and transform processes, and building an organized (mass) base of women – and men – with growing levels of political consciousness, engagement and agency, enhanced space, voice and visibility; the creation of new skills and capacities for women in the climate change arena; enhanced bodies of information and knowledge on gender and climate change; the claiming of rights and gaining of concrete new resources and assets for women; the reframing of existing discourses on climate change and its social implications; changes in power relations in the climate change arena; sensitization of other social movements; and increased public awareness and sensitization on gender aspects of climate change. Although still in process, many of these elements are in line with what Roa and Kelleher (2005) identified as the necessary 'dynamics of change'.

As IUCN's senior gender advisor Lorena Aguilar (2002) argues: sustainable development is not possible without gender equity and equality. This notion also asks for caution: we should not see women as merely instrumental to climate change action, otherwise women's civic activism would only add to their already existing burdens of unpaid caring responsibilities and paid employment and pertain traditional gender roles (MacGregor, 2006). It is crucial that women, like men and children, also benefit directly from the shifts in paradigms in climate change policies and actions. And women's leadership is an important element in advancing equity and social justice for all (Batliwala and Rao, 2002). This chapter described and analysed some of the most visible manifestations of that leadership, but is by no means complete in its description: around the globe thousands more local, national and regional initiatives exist that reflect women's leadership in the environmental and climate change arena. These can be instrumental in enhancing women's participation in environment and climate change policy-making, decisions and initiatives, and in promoting that women's priorities become central in all environment and climate change policies and actions. At the end we need partnership power (Eisler and Corral, 2007). It are the male and female policy-makers, planners and technical experts themselves that must be willing to share their power, and to reframe their climate change policies and programmes.

References

Aguilar, L, Castañeda, I. and Salazar, H. (2002) *In Search of the Lost Gender. Equity in Protected Areas*, IUCN-Mesoamerica, San José

Arts, B., Dankelman, I. and Rijniers, J. (2002) 'Duurzame ontwikkeling: Dimensies, strategieën, valkuilen en perspectieven', in B. Arts, P. Hebinck and T. van Naerssen (eds) *Voorheen de Derde Wereld: Ontwikkeling anders gedacht*, Metz, Amsterdam, pp37–60

Batliwala, S. (2008) *Changing Their World – Concepts and Practices of Women's Movements: Building Feminist Movements and Organizations*, report, Association of Women in Development, Toronto/Mexico DF/Cape Town

Batliwala, S. and Rao, A. (2002) *Conversations with Women on Leadership and Social Transformation*, report, Gender at Work, Toronto

Bishnoi Organisation (2010) *Saka: Self Sacrifice for the Cause*, www.bishnoi.org/khejadli.php, accessed 15 February 2010

Both ENDS (2010) *Joke Waller-Hunter*, www.bothends.org/index.php?page=2_2_2, accessed 20 March 2010.

Brownhil, L. and Turner, T. (2006) *Climate Change and Nigerian Women's Gift to Humanity*, Centre for Civil Society, University of KwaZulu-Natal, Durban

Dankelman, I. and Davidson, J. (1988) *Women and the Environment in the Third World: Alliance for the Future*, Earthscan, London

Eisler, R. and Corral, T. (2007) 'Leaders forging change: partnership power in the 21st century', in W. Link, T. Corral and M. Gerzon (eds) *Leadership Is Global: Co-creating a More Humane and Sustainable World*, Shinnyo-en Foundation, San Francisco, pp65–79

Hoppe, R. (2010) 'Lost in translation? Boundary work in making climate change governable', in P. Driessen, P. Leroy and W. van Vierssen (eds) *From Climate Change to Social Change: Perspectives on Science-Policy Interactions*, International Books, Utrecht, pp109–130

ISDR (International Strategy for Disaster Reduction) (2008) *Gender Perspectives: Integrating Disaster Risk Reduction into Climate Change Adaptation: Good Practices and Lessons Learned*, UN International Strategy for Disaster Reduction, Geneva

Kitakyushy Forum on Asian Women (1993) 'Environment, development and women', Special Issue, *Journal of Asian Women's Studies*, vol 2, 1993

Maathai, W. (2003) *The Green Belt Movement: Sharing the Approach and the Experience*, Lantern Press, New York

Maathai, W. (2006) *Unbowed: A Memoir*, Alfred Knopf, New York

MacGregor, S. (2006) *Beyond Mothering Earth: Ecological Citizenship and the Politics of Care*, University of British Colombia Press, Vancouver

Macha, C. (2010) 'Active involvement of women in environment decision making key', *Times of Zambia*, 6 March 2010

Pietilä, H. (2002) *Engendering the Global Agenda*, development dossier, UN Non-governmental Liaison Service, Geneva

Roa, A. and Kelleher, D (2005) 'Is there life after gender mainstreaming?', *Gender and Development*, vol 13, no 2, pp57–69

UNEP (UN Environment Programme) (2004) *Women and the Environment*, UNEP, Nairobi

UNEP (2005) *GEO Yearbook 2004/5*, UNEP, Nairobi

WECF (Women in Europe for a Common Future) (2002) 'Why women are essential for sustainable development. Results of the European Conference for a Sustainable Future, Čelákovice (Prague), 14–17 March 2002', WECF, Munich/Bunnik

WECF (2009) 'Women in Europe for a common future', brochure, WECF, Munich/Utrecht

Websites

ENERGIA, www.energia.org
Gender CC, www.gendercc.net
GGCA, www.gender-climate.org
IUCN, www.iucn.org/about/work/programmes/gender/
WECF, www.wecf.eu
WEDO, www.wedo.org
WOCAN, www.wocan.org

CASE STUDY 9.1

CLIMATE JUSTICE THROUGH ENERGY AND GENDER JUSTICE: STRENGTHENING GENDER EQUALITY IN ACCESSING SUSTAINABLE ENERGY IN THE EECCA REGION

Sabine Bock, Gero Fedtke and Sascha Gabizon

Introduction

Throughout the Eastern Europe, Caucasus and Central Asia (EECCA) region energy supply is a problem, especially in rural areas. Households rely to a great extent on fossil fuels or biomass – dried dung, shrubs and firewood, but they also use all kinds of waste, such as plastics. This has a severe impact on the environment, contributing considerably to deforestation and land degradation, which in turn has adverse effects on the quality of life of the villages' population. Air pollution created by plastics, tar and other types of fuel leads to serious health problems, including respiratory illnesses. Women and children are the most affected by indoor pollution and, according to the World Health Organization (WHO), 21,000 people die of indoor air pollution every year in the EECCA region. Insufficient availability of warm water for hygienic purposes likewise adversely affects human health. A major problem is providing warmth during the often cold winters; and daily routines such as cooking or heating water for bath or laundry put a stress on both the environment and the villagers' budgets as well. Electricity is also a problem with blackouts and power cuts, even in the capitals. And energy prices are rising rapidly.

The EECCA region is already affected by climatic changes. This manifests itself, for example, in melting glaciers, increased aridness and desertification; and access to water is becoming a major challenge. These effects are enhanced by human behaviour, while policy frameworks are not effectively managing climate change adaptation and mitigation. Energy efficiency and renewable energy technologies are not well-known and, under current conditions, are not really affordable. International financial mechanisms under the United Nations Framework Convention on Climate Change (UNFCCC), such as the Clean Development Mechanism (CDM), are not accessible for local communities. At the same time, people are not aware of the possibilities of sustainable energy sources and the chances they provide.

The role of women

Women bear the brunt of these negative effects; and they are key actors in managing the difficulties in everyday life. Large-scale labour migration of men due to a lack of income-generating opportunities further contributes to this effect. Village people hardly see opportunities for development in their own region. At the same time, women's perspectives and their roles are not adequately reflected in hierarchic village power structures, religious institutions, nor in state policies and many international projects. Such structures can be barriers to positive developments in villages. Also, international aid programmes sometimes contribute to a passive mentality by accustoming people to free supplies of goods and modern technologies. But instead of such negative developments, people could become sustainable energy managers themselves, based on their own needs and skills.

The work of the WECF network

Women in Europe for a Common Future (WECF) and its partners focus on capacity building of local non-governmental organizations (NGOs) and stakeholders. A main objective is to provide people with access to sustainable energy solutions for rural areas and to contribute to achieving the internationally agreed Millennium Development Goals (MDGs), particularly MDG7 (ensuring environmental sustainability), MDG1 (reducing extreme poverty) and MDG3 (empowering women and strengthen their participation in decision-making processes). WECF's approach is actor-oriented and participatory, with expressed local needs being the basis for supporting community action. Cross-cutting issues, particularly environmental quality, gender equality, human rights and the inclusion of marginalized groups, are core in implementing projects.

WECF focuses on identifying women's and men's specific knowledge and skills in the production, maintenance and use of sustainable energy resources and technologies. The organization tries to build on existing skills of women and to decrease their time and work load. In one project, for example, women are producing traditional insulation materials in order to have a share of the new and upcoming energy market. The production of fruit solar driers gives women the possibility to generate income. And the installation of hot water boilers improves the quality of everyday life for women in general.

Project example

The WECF project 'Climate protection through sustainable energy by training, capacity building and networking of local NGO partners in the Caucasus and Central Asia' is supported by the International Climate Protection Initiative (that is based on a decision of the German Parliament, by the Federal Ministry for the Environment, Nature Conservation and Nuclear Safety). Three training programmes were conducted for WECF partners and other NGOs, representatives of authorities, the private sector and scientists. The project focused on the exchange of knowledge and experiences on climate change, sustainable energy initiatives and the implementation of local sustainable energy supply technologies, both from a theoretical and practical perspective. The content of the trainings was based on local needs and experiences, such as solar energy and biogas in Georgia, for example, or insulation and solar collectors in Kyrgyzstan.

The three training sessions of one week each took place in different countries and regions of the world: in Munich (Germany), Tbilisi (Georgia) and Bishkek (Kyrgyzstan). Furthermore, participants had possibilities to participate in international conferences: the UN climate change conferences in Bonn (Germany) and Poznan (Poland). Thus this provided a combination of training and direct participation. The focus at the training sessions was on technology and knowledge transfer on different levels – from implementation to policy issues and the international UNFCCC process. It was important to show the diversity of sustainable energy options, from low cost to more advanced, for example, but also more expensive solutions. The programme was a combination of presentations, excursions, workshops, discussions, conference days and practical implementation. Examples of these are the construction of solar collectors and fruit driers in Armenia and in Georgia, insulation for a school in Azerbaijan and a feasibility study on micro-finance systems for biogas and solar installations in rural areas in Georgia. At the Conference of Parties (COP-14) of the UNFCCC, December 2008 in Poznan, a delegation of the partner NGOs from the EECCA region presented proven successes to reduce energy poverty by sustainable energy and they shared the experiences from the region in a side event. There they discussed how to bridge the gap between policy and practice.

Conclusions

Improved knowledge and capacity of the trainees, especially of the 18 partner NGOs of WECF, is the most important result of this initiative. This was made visible in demonstration projects, policy recommendation papers and presentations at national and international level. The built-up knowledge and capacity will have further impacts, in particular in social, economic and environmental areas. Sustainable energy can be a way out of energy poverty and towards independence from imported energy. Saving money for energy means more income for other needs. Reliable energy supply means better living and working conditions. It also improves health conditions, especially of women and children, with activities such as floor insulation and reduced indoor pollution. Saving energy by using, for example, energy efficient stoves, will have an important effect on the ecosystem by mitigating environmental degradation, deforestation and greenhouse gas (GHG) emissions.

Case Study 9.2

National Federation of Women's Institutes: Women Organizing for a Healthy Climate

Ruth Bond and Emily Cleevely

The UK, along with the rest of the world, has a duty to future generations to address and reduce its climate change impacts. It can do this by both reversing its status as a prolific emitter and also providing technology, knowledge and funding to developing nations to enable them to continue their pattern of growth without a corresponding growth in emissions.

In the future, the UK will also be affected by the ravages of climate change, with low-lying areas being lost to sea level rises, increasingly unpredictable weather patterns and more frequent incidents of severe flooding. Arguably, the weather is already becoming erratic and demonstrates the effect our carbon-intensive behaviour is having on the environment.

Much-needed steps have been taken in the UK in recent years to tackle the inequality that women suffer in areas such as equal pay, public office and access to jobs, goods and services. However, legislation can only do so much and there is a long way to go before we make the cultural change needed in our work places and public life, to truly overcome gender inequality.

Women can feel shut out of the climate change debate, which is male-dominated. Yet, climate change is an area where women are uniquely placed to make a real difference in their homes and communities.

In the UK women are responsible for around £400 million domestic expenditure every week (Visa, 2007) and with women more conscious of their environmental impacts than men, this provides a fantastic opportunity for every woman in the UK to make consumer choices which are the best possible for the environment. It is these seemingly small personal decisions which the National Federation of Women's Institutes (NFWI) has tried to help its members across the

UK build upon to make a big difference nationally and demonstrate that ordinary people can make a huge impact when they act together. The NFWI is the largest organization for women in the UK, with 205,000 individual members.

As part of its Women Reaching Women project (funded by the Department for International Development and run jointly with Oxfam UK and the Everyone Foundation), NFWI members across England and Wales have been finding out about the ways in which women around the world are affected by the changing climate. Discovering more about the challenges facing women in some of the most vulnerable areas of the world, as well as the response of women in other nations, has led NFWI members to understand both the vulnerability and the potential of women all over the world. Education among the NFWI membership and, in turn, individuals passing on information to their own communities, has ensured that awareness of the issues has increased and deepened.

The NFWI, as part of the UK's Stop Climate Chaos Coalition, was instrumental in lobbying for the UK's Climate Act which sets the first legally binding carbon reduction targets in the world. Throughout 2009, the NFWI ran a campaign which focused on increasing awareness amongst members and their communities, as well as amongst politicians and other non-governmental organizations (NGOs) on the issue of Women and Climate Change. This had real impact in the lead up to the Conference of Parties (COP-15) in Copenhagen in December 2009. Yet there is still more to be done to convince decision-makers and ordinary people alike that there is a unique link between women and climate change.

The NFWI has responded to the climate challenges faced by women by 'doing' – saving energy, saving resources, raising awareness, delivering its views to the UK government and taking part in The Wave (the UK's biggest ever climate change demonstration, attended by 50,000 people in December 2009). Throughout the history of NFWI campaigning on climate change, the organization has seen that it is important to demonstrate action on the ground which inspires others and provides politicians with the mandate they need to bring about broad-reaching national action on the issue. Some of its practical projects have included the Carbon Challenge, in which 20,000 participating members saved 20 per cent of their baseline emissions over the course of the project; the Eco Teams project, in which NFWI members received training to go out into their communities and achieve tangible savings in energy usage, waste and transport; and the Love Food Hate Waste campaign which saw members working in food teams to learn more about how to reduce waste from food. Taking strong action within communities has meant that the NFWI is one of the UK Government's valued stakeholders, particularly for the Department of Energy and Climate Change.

In light of the failures of the United Nations Framework Convention on Climate Change (UNFCCC) process at Copenhagen, 2010 sees the NFWI redoubling its efforts in grassroots action. Without leadership from international governments and the UNFCCC process, it is down to the mass of ordinary people all around the world – and women in particular – to demonstrate the desire and determination to change the way they live. By making sweeping changes in their everyday lives, women can start to have a real impact and also prepare themselves for the changes that will be possible when governments drive forward a new agenda from above. We know that, by working together within their communities, women can be a powerful voice for change.

Reference

Visa (2007) 'Visa Europe Report', 11 July 2007, www.visaeurope.com/pressandmedia/newsreleases/press313_pressreleases.jsp, accessed 10 February 2010

Case Study 9.3

Women and the Environmental Justice Movement in the Niger Delta Region of Nigeria

Omoyemen Odigie-Emmanuel

The Niger Delta is one of the world's largest wetlands and consists of virtually all coastal states in Nigeria, namely Rivers, Delta, Bayelsa, Cross River, Akwa-Ibom, Edo, Abia, Imo and Ondo states. The Niger Delta extends over about 70,000km^2, and some 31 million people of more than 40 ethnic groups live in the area (Folorunsho, 2009).

The Niger Delta is one of the world's richest areas in terms of natural resources. Apart from its substantial oil and gas deposits, there are extensive forests, abundant wildlife and fertile agricultural land. The delta is also famous for its fish resources (Okonta and Douglas, 2001). However, since oil was struck four decades ago, the ecological and environmental hazards from indiscriminate exploration and exploitation have constituted an affront on the community and the survival of its people (Akiyode-Afolai and Iyare, 2005).

The Niger Delta region is suffering from large-scale environmental and human rights abuses evident in the oil spills and gas flaring and their consequential destruction of water sources and vegetation, destruction of livelihoods and damage to food supply, ill health, increase in incidents of conflicts and sexual vulnerability and exploitation.

Events in the Niger Delta reveal that women, after years of suffering and death, have taken their place among the oppressed and begun to demand environmental and ecological justice in the region. According to Turner and Leigh, women were at the forefront of social movements because despite the fact that they are largely unwaged, capital exploits them as commodities and uses up 'free' nature, social services, building space, and the production of paid and unpaid work (Turner and Leigh, 2004).

Women protested and got involved in social movement actions, such as the Ogharafe women/youth collaborative occupation of the production site of Pan Ocean – a US multinational corporation – in

1984, and the Bonny women protest against Shell in March 1986. The Egi women organized protests and street marches against oil company Elf Aquitaine in September and November 1998, and in January 1999 the Niger Delta Women for Justice led by Annie Brisibe held protests. Noticeable are also the siege by Ugborode women at the Chevron facilities in Escravos on 8 July 2002, the occupation of Abiteye flow station on 15 July 2002 by Kenyagbene women and Ekpan women's protest against Chevron on the 30 July 2002.

The climax of women's involvement in the struggle for environmental and ecological justice is the united campaigns against two oil giants, Chevron and Shell (Okon, 2004). On 8 August 2002, women of Itsekiri, Ijaw and Ilaje peoples occupied the headquarters of these multinational companies protesting against degradation of their land and corporate irresponsibility.

The main reason for the action of the women was that they were weary of the degradation of their land and waters, the destruction of their livelihood, the economic and social underdevelopment of the Niger Delta. Women are suffering both direct and indirect consequences, including the feminization of poverty and their marginalization. They were tired of complaining and decided as agents of change to take action for their liberation and shaping their own future and that of their communities.

The implication of women's involvement in struggles for environmental and ecological justice seems to be that the social justice movement in Nigeria takes new shape in the future with women taking proactive steps and women's role reshaping the violent implications of conflict.

References

Akiyode-Afolabi, A. and Iyare, T. (2005) *The 11-Day Siege: Gains and Challenges of Women's Non Violent Struggle in Niger Delta*, Women's Advocates Research and Documentation Center, Nigeria

Folorunsho, R. (2009) 'Coastline erosion and implications for human and environmental security', presentation at the two-day International Conference on Climate Change and Human Security in Nigeria: Challenges and Prospects, Nigerian Institute of International Affairs, Nigeria

Okon, E. (2004) *Rural Women Under Siege*, Niger Delta Women for Justice, Nigeria

Okonta, I. and Douglas, O. (2001) *Where Vultures Feast: 40 Years of Shell in the Nigeria Delta*, Environmental Rights Action, Nigeria

Turner, T. and Leigh, B. (2004) *The Twenty-First Century Land And Oil Wars: African Women Confront Corporate Rule*, International Oil Working Group, New York

10

Epilogue: From Divergence towards Convergence

Irene Dankelman

The year 2010 commemorates 15 years since the Beijing Fourth UN Conference on Women, and is 25 years since the implications of environmental conditions for women's well-being were taken up at the World Conference to Review and Appraise the Achievements of the UN Decade for Women in Nairobi, 1985. The UN's *Forward-Looking Strategies for the Advancement of Women* mentions that the 'deprivation of traditional means of livelihood is most often a result of environmental degradation resulting from such natural and man-made disasters as droughts, floods, hurricanes, erosion, desertification, deforestation and inappropriate land use'(UN 1985, paragraph 224). So, although not brought up explicitly, the notion of climatic changes – with droughts, floods and hurricanes – was already clearly reflected in this outcome document.

Three years later, Joan Davidson and I developed the book *Women and Environment in the Third World: Alliance for the Future* (Dankelman and Davidson, 1988). And on 6 December that same year, the UN General Assembly agreed on Resolution 45/53 that outlined the mandate for the establishment – in 1989 – of the Intergovernmental Panel on Climate Change (IPCC) by the World Meteorological Organization (WMO) and the United Nations Environment Programme (UNEP). In 1989, Vandana Shiva also published her compelling work *Staying Alive: Women, Environment and Development* (Shiva, 1989).

Such developments all culminated in the United Nations Conference on Environment and Development (UNCED) or Earth Summit in Rio de Janeiro in 1992, its United Nations Framework Convention on Climate Change (UNFCCC) and its recognition of women's role in environmental management and sustainable development. It was in that same year that my co-author and friend Joan Davidson sadly passed away. The Fourth World Women's Conference in Beijing in 1995 and its Platform for Action brought the environmental context again to the agenda for enhancing gender equality. However, in the Millennium Development Goals

(MDGs) of the Millennium Summit in 2000, the reduction of poverty, the enhancement of gender equality and environmental sustainability were again presented as independent, separate goals.

Towards convergence

Now that we are a quarter of a century further in time and in future's history, it seems that, from a tendency to diversify development, human rights and environmental goals and objectives, a far more holistic vision of the need to see and recognize the interlinkages between all these relevant societal aims in our policies, programmes and actions has evolved. The convergence of the stream on environmental sustainability and the one on gender equality should result in better, more robust and more efficient climate change policies and actions, and into more equality, justice and safer livelihoods for women and men coming from all backgrounds.

In this context the meaning of the word 'convergence' seems to be very adequate: from a divergence or failure to conform or match – deviating us from our expressed wish – 'convergence' stands for the coming together from different directions united around one common goal. In biology – my personal background – convergence means the tendency of different species to develop similar characteristics in response to a set of environmental conditions – in this case climatic changes. For ophthalmologists it encompasses turning the eyes inward, and in meteorology it refers to the meeting of air masses. For me personally, beyond this semasiology, the meaning of 'convergence' became visible in the late 1980s when, during our field-work on agrobiodiversity and gender issues, I travelled with colleagues and friends through the north of India and was pointed to the convergence of two major rivers, the Ganga and the Varana, near Benaras, a place of sacredness. Similar sacred places of more rivers converging to one stream can be found all over the world, for example in Pittsburgh, US. There, the Point Park area, where three rivers converge, was a sacred site for tribal rituals of the Native Americans that inhabited that region (Hanchin, 2008). The coming together of many streams towards a common goal, that is how this final chapter got its title 'From Divergence towards Convergence'.

Reflection

In co-creating this book, we learned from many experiences in the world: we listened to the voices of female farmers and land managers, we looked into the lives of urban dwellers, particularly in the global south, and we reflected on the impact of male and female consumers in the global north. We analysed the impacts of climate change on the lives of women and men with their diverse backgrounds and in their specific contexts, and we learned about their resilience and coping strategies. We saw how national and international institutions struggle with mainstreaming gender into climate change policies and actions. We explored existing policy frameworks and their potentials for enhancing gender responsiveness in climate change work. And finally, we have discovered how women, along with men, have organized locally and at international level as agents of change for a healthy and just climate. Finally we have arrived at a stopover in our ongoing journey, at the convergence of two streams, ready to follow our path along in a more powerful, caring and beneficial way.

Summarizing, this book has taught us about:

- The different responsibilities, roles and knowledge systems of women and men with regard to natural resources use and management. And the inequalities in rights, livelihood perspectives and opportunities amongst both genders.
- The threat that the complex and wicked problem of climate change in this era will increase existing inequalities, particularly gender inequality.
- The pressure on people's resilience and coping strategies, and the urgent need for adaptation and mitigation activities. In this respect much can be learned from the existing gender-disaster literature.
- The observation that these situations manifest themselves already in many parts of the world, not only in rural areas but also in urban situations. And the need to work in a context-specific way.
- The fact that it is most often – but not always – women who feel the consequences of climatic changes in their livelihoods and lives.
- The intersectionality of climate change, that asks for an integrated approach at all levels, strengthening interdisciplinarity as well as transdisciplinary cooperation among different sectors of society.
- The numerous necessities and opportunities to make climate change adaptation and mitigation more gender-responsive and to enhance gender equality in climate policies and actions. And the warning that climate safety and gender justice do not always coincide automatically.
- Mitigation systems that limit greenhouse gas (GHG) emissions and promote gender equality, through technological innovations changes in lifestyles, and reallocation of investments.
- The fact that the policy environment and society as a whole are just starting to become aware of these interfaces and nexuses; but that our

institutions and programmes are still very sectorally organized and do not appropriately reflect these insights. They still lack the social changes needed (Driessen et al, 2010).

- Legal policy frameworks that offer chances to work out a more comprehensive approach towards engendering climate change; but on the other hand many of these are actually still disconnected.
- And last but not least, regarding women's multiple roles as agents of change, organizing, adapting and making more sustainable consumption and production choices, promoting a more holistic approach and the need to involve men.

Challenges ahead

In its introductory mode, this book is still limited in scope, extent and depth. Working on gender and climate change challenges us to look deeper into our societies. Maybe it leaves us with more questions than solutions but in order to adequately tackle the issues of gender inequality and climate change, these questions have to be raised. We still need to ask ourselves what has caused all the global problems we are facing and analyse why the present economic, financial and political systems have failed in preventing major local and global social and ecological disasters. We have to look into new ways to delve deep into the premises, frameworks and paradigms of our current economic, social and political systems, from a human rights and eco-sane perspective. In this context we could learn much from scholars, such as Hazel Henderson (1996), Maria Mies (1998), Ariel Salleh (2009) and Mary Mellor (2010), who have written extensively about eco-feminist political economy as an alternative for contemporary economic systems. Building such a basic understanding and awareness, needs the contributions and interdisciplinary and transdisciplinary cooperation of many open-minded experts, people with an academic background and those who live their lives in the complex realities of today, women and men alike.

Secondly we still have to learn how to deal with the inherent uncertainties and complexities that are linked to global environmental challenges, in particular climate change, and human behaviour. The ability to plan and act with precaution, can guide us in that 'unknown region'. We have to find ways to limit our human footprint, realizing that we are living in a globalized world – full of chains of interaction in space and time, and related impacts for present and the future generations. On the other hand, there is already much knowledge and wisdom on how to fully take into account the characteristics, flexibilities, potentials and limitations of our world and its peoples.

In our focus on gender issues in the context of climate change, we have learned about the persistent inequalities in human society and how these hinder and obstruct climate justice and equal rights to ecological services. Why does it seem that our human mindset, organization and geo-political system are inadequate to deal with basic human rights, with safeguarding a descent live for all, women and men in all localities alike. Is it human psyche, our culture, our lust for power and mimetic desire that push us in ever more demanding consumption and production systems (Girard, 1961; Achterhuis, 1988)? Is it the focus on personal or family profit and well-being instead of that of the local and global community? The tragedy of the commons in which shared goods clash with our desire for sovereignty and independence? Is there a way to enhance our 'caring capacity'?

Medha Patkar, social activist and founder of the Nardama Bachao Andolan organization that resisted the damming of the Narmada River in India, summed things up as follows:

> *If the vast majority of our population is to be fed and clothed, then a balanced vision with our own priorities in place of the Western models is a must. There is no other way but to redefine 'modernity' and the goal of development, to widen it to a sustainable, just society based on harmonious, non-exploitative relationships between human beings and between people and nature.* (www.rightlivelihood.org/narmada.html, accessed 14 January 2010)

Raising such fundamental questions can leave us overwhelmed, disconcerted and with a feeling of apathy. However, the many experiences in this book show that there is a lot that moves well, and that there are many initiatives, at local and global level, that countervail the above-mentioned dilemmas, strengthen positions of women and men, and contribute to climate change mitigation and adaptation. Not only should we learn from these initiatives, but also look into ways to strengthen, upscale and enhance these and to promote a favourable policy and institutional environment.

This is not only a question of empowering and mobilizing women, but also of looking much more into the masculine aspects of gender and climate change and into the roles that men can play in these processes. Being open-minded and inviting men as crucial companions in this common effort.

Strategic steps

In this book a wide range of strategic steps are proposed:

Towards more holistic approaches

There is a need to shift the focus of discussion on climate change away from a primarily technocratic exercise to one employing the language of global justice and human rights, including the right to development and gender equality and equity. Boundary work can help in bridging the different spheres and sectors. Climate change responses and strategies should build on women's local and indigenous realities and knowledge systems, combined with technological and societal innovation. As it is actually the quality of relationships that determine the quality of a complex system, there is an urgent need to link local realities with national and global decision-making.

Apply gender and power analyses

The understanding of the interface between the socio-sphere and eco-sphere and of gender aspects of environmental and climate change could be significantly enhanced by application of gender analyses. The causes of vulnerability and capacity have to be analysed clearly. Gender analysis has to be completed with an understanding of other intersecting identities that can marginalize people: an intersectional analysis can help build a textured understanding of power, gender and intersectionality (Harding, 2004; Yuval-Davis, 2006).

Build understanding and knowledge

Through an intensified effort to document and study observations and experiences of local women and men with regard to climatic changes, the common understanding and knowledge base on gender and climate should be strengthened. Develop and support a comprehensive international research agenda, in which academia and institutions from around the globe cooperate. Explore in these also relatively 'new' areas, such as gender, health and climate change (WHO, 2010). There is an urgent need to learn from local indigenous practices and to recognize the potential contribution of local and indigenous women and men in climate change mitigation and adaptation. All societal climate change statistics should disaggregate data by gender (see Case study 10.1).

There is a great need to document more mountain women's lives and the gender issues common to the Himalayas. Little research has been pursued on mountain populations, particularly how they adapt to change, and how gender-specific conditions affect their abilities to adapt. (Leduc et al, 2008 p8)

Enhance capacity and expertise on climate change and gender

There is an urgent need to raise awareness on climate change and gender equality, through data collection, documenting, dialoguing, education (curricula) and the use of media. Develop a roster of gender and climate change experts and further improve the research base.

Promote participation and access to information

Improve women's participation in decision-making on climate change and draw on skills and agencies of women and marginalized groups. Guarantee the representation of poor women and men in local and sub-regional formal decision-making structures and the active and gender balanced representation of indigenous peoples. Support women to develop voice and political capital to demand access to climate change decision-making, mechanisms and risk management instruments. Promote the active participation of women's organizations. As gender discrimination in access to information and institutional support affects women's vulnerability and participation in climate change decision-making and benefit-sharing, there is an urgent need to improve women's access to climate information (see Case study 10.2).

Reframe policies and mainstream gender

In order to prevent climate change and energy policies and actions actually reinforcing disparities between men and women, it is vital to consider the gendered effects of policies. Gender equality has to be mainstreamed in all climate change policies and actions, and gender provisions have to be included in international policy instruments. In pro-poor climate just deal(s), gender equity has to featured prominently. There is already a wide range of legal gender-related instruments (soft and hard law) that can inform emerging policies and practices on climate change. Pro-poor and gender responsive policies and strategies have to help households stabilize consumption and push the private sector, including the transport, energy and agricultural sectors, to climate-proof transitions.

Box 10.1 *Gender audit of energy policy in Botswana*

The Botswana Technology Centre (BOTEC), in consultation with the Energy Affairs Division of the Ministry of Minerals, Energy and Water Resources and other stakeholders, executed a gender audit of Botswana's national energy policies. Botswana is the first country where such an audit was held. The audit showed that although there is a common understanding of the different roles of women and men in Botswana, the knowledge on the relationship between gender, energy and poverty was still limited. This has resulted in gender-blindness of the energy policies and programmes, and a lack of consultation with household residents and women in particular, in developing the energy policy. The audit also showed a lack of gender-disaggregated data and a general lack of association between energy services and the MDGs. Based on this audit and follow-up trainings, the awareness in the government and of the Botswana Power Corporation (BPC) staff has increased. BPC recently started a groundbreaking gender mainstreaming programme for their rural electrification programme. The audit also led to a pilot project for collecting gender-disaggregated data, and to strengthening gender expertise in the country's energy sector.

Source: Wright and Gueye (2009)

Review planning

Consider land-use planning, including urban planning and climate change preparedness plans from a poverty and gender responsive lens. Through such planning local capacity to cope with, adapt to and mitigate climate change should be enhanced. Integrate pro-poor and gender-sensitive criteria into planning, design, implementation, monitoring and evaluation of climate policies and actions, and support local community action plans for adaptation and GHG mitigation and reduction of climate related hazards.

Strengthen resilience

There is not only a need to promote socio-ecological resilience, coping strategies, adaptation capacity and improve disaster preparedness and management, but this needs to be gender-sensitive and pro-poor. Resources and assets of local communities, particularly women and those living in poverty, need to be strengthened. Improve people's health in their communities and work places as important components of ensuring resilience. Individuals and households have to be enabled to develop sustainable and resilient livelihoods, for example through diversifying income and strengthening economic base. Early warning systems and climate information should be made available and accessible at a local level (see also Case study 10.2).

Support adaptation capacity

Addressing climatic changes could create opportunities for achieving social goals, but inequalities may be made worse without gender analysis and gender sensitivity. Therefore strengthen gender-sensitive and pro-poor adaptation strategies, dependent on availability of resources and assets, and focusing on the most vulnerable areas and groups. Planning for adaptation to long term change must be founded on women's and men's knowledge and experiences as they make choices in an uncertain climate. This should also be reflected in the National Adaptation Programmes for Action (NAPAs) (UNFPA and WEDO, 2009). Gender-based preferences for livelihood strategies in response to long-term change should have implications for adaptation decisions. As women's assets largely determine how they will respond to impacts of climate change, actions should build up the asset base of women and therefore their capacity to adapt (Rodenberg, 2009).

Promote gender responsive mitigation

Mitigation of climate change should have benefits for poverty reduction and women's empowerment. Ensure that women have a voice in developing mitigation strategies and enable their choices as consumers. Involve women in clean technology development and ensure their access to technologies. Women should also be trained in cases of technology transfer. Invest in energy infrastructure technologies and end-uses that directly meet women's energy needs and make their labour more productive, for example, through improved cooking stoves and fuels, food processing technologies, drinking water pumping and transport, electric lighting and media (Oparaocha, 2009). Make sure that local women benefit from Reducing Emissions from Deforestation and Forest Degradation (REDD) mechanisms. Strengthen the notion of 'eco-sufficiency and global justice' (Salleh, 2009).

Encourage gender-sensitive financial mechanisms and instruments

It is important to understand the opportunities and dangers for gender equality presented by new financial mechanisms. They can significantly contribute to the empowerment of women or, on the contrary, deepen the already existing gender gap. There is an urgent need to develop a set of gender-sensitive criteria for all new and existing climate change financing mechanisms supporting adaptation, mitigation, capacity building, technology cooperation. Allocate adequate resources to address the need of women in climate change, and make adaptation funds accessible to local and indigenous women (Peralta, 2008; Schalatek, 2009).

Promote women's empowerment and organization

Promote the empowerment of women and marginalized groups as central elements of an equitable response to climate change by improving their access to skills, education and knowledge. Build women's climate change capacities in rural, urban and peri-urban areas, and enhance their eco-literacy and leadership (Link et al, 2007). Educate women on climate to enable them to effectively play their role as agents of change in mitigation

Box 10.2 *Women's Green Business Initiative*

The Women's Green Business Initiative to promote women's entrepreneurship opportunities in climate change adaptation and mitigation, has recently been launched by the United Nations Development Programme (UNDP) and partners in the Global Gender and Climate Alliance. The initiative works closely with women in developing countries to start, incubate or scale-up business enterprises, fostering their entrepreneurial capacity to access climate funds in a new green economy. By promoting business activities managed by women, it intends to help them to establish sustainable livelihoods, to protect critical ecosystems and to strengthen communities' resilience to climate change threats. A practical example of how local women can access climate funding to finance green business activities for environmental protection and income generation is the local women's group Twiyubake Turengera Ibidukikije ('Empowering Ourselves in the Protection of the Environment') in Rwanda. In 2008, the UNDP Gender Team offset the GHG from their travel to a meeting in Rwanda by providing a voluntary credit grant of US$10,420 to the local women's group for a bamboo-growing project. The women used the grant to plant 60ha of bamboo trees for use in making furniture, baskets and handicrafts. Besides offsetting the GHG emissions from the meeting, the project is reducing deforestation and erosion in the Nyungwe Forest National Park, and offers the women a stable income.

Source: UNDP (2009)

and adaptation. As a wide range of local, national and international women's organizations are active in this area and act as important catalysts in empowering women, developing own analysis and bringing gender perspective to policy level and development practices, involve and support these organizations.

Conclusions

It is clear that there is a need for transformative change. In that sense the conclusion of gender expert and advocate Srilatha Batliwala (2008, p11), while analysing feminism today, is also relevant in the context of gender and climate change:

> *We now stand not only for gender equality, but for the transformation of all social relations that oppress, exploit, or marginalize any set of people, women and men... We seek a transformation that would create gender equality within an entirely new social order, one in which both men and women can individually and collectively live as human beings in societies built on social and economic equality, enjoy the full range of rights, live in harmony with the natural world, and are liberated from violence, conflict and militarization.*

From a divergence of interests and perspectives that results in more inequality and increasing problems related to climate change, we urgently need to move towards the convergence of environmental and societal aims, goals, strategies and approaches in our institutions, policies and economies. The convergence of both streams should result in climate and gender justice, and reframe the way we stand in and deal with the Earth and human society in their diversity and totality.

We started this book as a journey with the testimony from Rehana Khilji in Baluchistan, Pakistan; we finish our journey at the point where the streams of gender and climate change converge, the place that is holy in many cultures; forming a river of opportunities ahead of us – co-creating the future along its banks and shores.

References

Achterhuis, H. (1988) *Het Rijk van de Schaarste:Van Thomas Hobbes tot Michel Foucault* (The Realm of Scarcity: From Thomas Hobbes to Michel Foucault), Ambo, Baarn

Dankelman, I. and Davidson, J. (1988) *Women and Environment in the Third World: Alliance for the Future*, Earthscan, London

Driessen, P. J., Leroy, P. and van Vierssen, W. (eds) (2010) *From Climate Change to Social Change: Perspectives on Science-policy Interactions*, International Books, Utrecht

Girard, R. (1961) *Deceit, Desire and the Novel: Self and Other in Literacy Structure*, John Hopkins University Press, Baltimore

Hanchin, V. (2008) 'Pittsburgh: Three converging rivers cortex', www.warriormatrix. com/about4783.html@sid=B533a2832bd22a8baeebcfdd59330ca2, accessed 22 February 2010

Harding, S. (2004) *The Feminist Standpoint Theory Reader: Intellectual and Political Controversies*, Routledge, New York

Henderson, H. (1996) *Creating Alternative Futures: The End of Economics*, Kumerian Press, West Hartford

Leduc, B., Shrestha, A. and Bhattarai, B. (2008) *Gender and Climate Change in the Hindu Kush Himalayas of Nepal*, International Centre for Integrated Mountain Development (ICIMOD) for Women's Environment & Development Organization (WEDO), Kathmandu/New York

Link, W., Corral, T. and Gerzon, M. (eds) (2007) *Leadership is Global: Co-creating a More Humane and Sustainable World*, Shinnyo-en Foundation, San Francisco, CA, www.sef.org

Mellor, M. (2010) *The Future of Money: From Financial Crisis to Public Resource*, Pluto Press, London

Mies, M. (1998) *Patriarchy and the Accumulation on a World Scale*, Zed Books, London

Oparaocha, S. (2009) 'Viewpoints: Interview with Sheila Oparachao, ENERGIA Network Coordinator/ENERGIA Programma Manager', *Boiling Point*, no 57, pp16–17

Peralta, A. (2008) *Gender and Climate Change Finance: A Case Study from the Philippines*, WEDO/Heinrich Böll Foundation North America, New York/Washington, DC

Rodenberg, B. (2009) *Climate Change Adaptation from a Gender Perspective: A Cross-cutting Analysis of Development-policy Instruments*, discussion paper no 24/2009, DIE (German Development Institute), Bonn

Salleh, A. (ed) (2009) *Eco-Sufficiency and Global Justice*, Pluto Press, London

Schalatek, L. (2009) *Gender and Climate Finance: Double Mainstreaming for Sustainable Development*, Heinrich Böll Foundation North America, Washington, DC

Shiva, V. (1989) *Staying Alive: Women, Environment and Development*, Zed Books, London

UN (1985) 'Forward-looking strategies for the advancement of women', UN, New York, www.un.org/womenwatch/confer/nfls, accessed 22 February 2010

UNDP (United Nations Development Programme) (2009) *Women's Green Business Initiative: Turning the Climate Change Challenge into Economic Opportunities for Women*, brochure, UNDP, New York

UNFPA and WEDO (United Nations Population Fund and Women's Environment & Development Organization) (2009) *Climate Change Connections: Gender, Population and Climate Change*, UNFPA and WEDO, New York

WHO (World Health Organization) (2010) 'Gender, climate change and health', draft discussion paper, WHO, Geneva

Wright, N. and Gueye, Y. D. (2009) 'Gender audits of energy policy in Botswana and Senegal: What has been achieved?', *Boiling Point*, no 57, pp9–11

Yuval-Davis, N. (2006) 'Intersectionality and feminist politics', *European Journal of Women's Studies*, vol 13, no 3, pp193–209

Case Study 10.1

Gender-disaggregated Data for Assessing the Impact of Climate Change[1]

*Ashbindu Singh, Jenny Svensson and
Janet Kabeberi-Macharia*

Introduction

It is unlikely that the impacts of climate change will be gender neutral, as its effects will increase the risk for women – the most vulnerable and less empowered groups in our society.

Without a gender-sensitive method of analysis, it is impossible to establish a full set of causes and potential effects of climate change. However, there is an overall lack of gender-disaggregated data in this area. In 2009 a study was executed by the United Nations Environment Programme's (UNEP) Division of Early Warning and Assessment – North America in order to analyse gender-disaggregated data with regard to climate change.

In this study, earlier research of the Organisation for Economic Co-operation and Development (OECD) 'Ranking port cities with high exposure and vulnerability to climate extremes' (OECD, 2007) – that focuses on port cities around the world with more then one million inhabitants (2005) – was used. A second set of data applied by is based on statistics from the United Nations Development Programme (UNDP), *Human Development Report 2007-2008* and from the *United Nations Statistical Commission* (UNSD, 2007). Rather than 'statistics on women', statistics on both women and men has been used to outline basic differences.

Existing gender-disaggregated data

The lack of gender-disaggregated data made the study challenging. Gender statistics is a relatively new field and cuts across all traditional fields (UNESCO, 2007). Rather than 'statistics on women', gender statistics are statistics on both women and men. It is thus equally important to highlight statistics concerning both genders. According to the United Nations Statistics Division (UNSD) and the Department of Economic and Social Affairs there has been limited progress in the gender specific reporting of official national statistics worldwide (UNESCO, 2007). Because of this insufficiency, we cannot fully understand how men and women will be affected differently in terms of climate change. However, the UNEP study provides an insight into some of the basic differences between men and women, and at the same time gives an indication on how to move forward in this debate.

There are categories in which there are sufficient gender-disaggregated data, but in others gender statistics are insufficient. In Figure 10.1.1 these categories are reflected. The factors that are outlined to have 'Insufficient Data' are a good indicator on where more research is needed within this field of study.

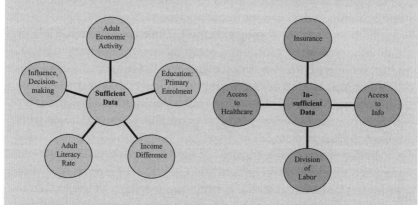

Figure 10.1.1 *Gender-disaggregated data (women/men): Sufficient and insufficient data*

The categories with sufficient information have been used in the UNEP study. All of the factors outlined and discussed below are considered to be of individual importance by themselves, and combined, specifically, they provide an understanding of how climate change will affect men and women differently in terms of climate change.

Categories with sufficient gender specific data include: adult literacy, influence and decision-making, education and primary enrolment, adult economic activity and income differences.

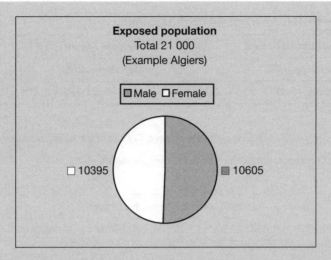

Figure 10.1.2 *Exposed population – without incorporating gender differences*

Creation of gender-disaggregated data

As a case study the UNEP analysis on the city of Algiers in Algeria is summarized here. When we assume that changes in climate conditions would affect men and women equally, the exposure is relatively evenly distributed between men and women. As shown in Figure 10.1.2 out of the 21,000 exposed people in Algiers 10,395 (49.5 per cent of the total population) are women, and 10,605 (50.5 per cent of the total population) are men.

However, when incorporating underlying social, political and economic factors, the exposure rate found in Figure 10.1.2 shifts significantly. Table 10.1.1 indicates how the exposure rate changes when underlying differences between men and women in literacy are considered.

Category 1: Adult illiteracy

Upon considering literacy rates in the case of Algiers, the underlying difference between women and men is evident at first glance. The percentage of illiterate women is, with 40 per cent, double that of men.

When considering illiteracy rates as an indicator for greater exposure to climate change impacts, we can clearly see a shift

Table 10.1.1 *Male and female illiteracy rate*

Male illiteracy rate	Female illiteracy rate
Percentage: 0.20	*Percentage:* 0.40
Illiterate: 10,605 × 0.20 = 2121	*Illiterate:* 10,395 × 0.40 = 4158
Of the 21,000 exposed people in Algiers: 6279 (2121 + 4158) are illiterate. 6279 is representing the whole illiterate population	
(total illiterate (men, female) / whole illiterate population)	
2121/6279 = (0.338) 33.8%	4158/6279 = (0.662) 66.2%

towards higher exposure of females. Previously the exposure rate was 1.0202 (10,605 men/10,395 women). In other words, if we are not to consider underlying differences between gender, men would be more exposed to changes in climate conditions than women since there are proportionally more men than women in Algiers. However, when we consider illiteracy as an underlying factor that affects peoples ability to adapt to changes in climate conditions the exposure rate is 0.510 (0.338 men/0.662 women). Men are now approximately only half as exposed to changes in climate conditions compared to women. Figure 10.1.3 illustrates the different exposure rates between men and women.

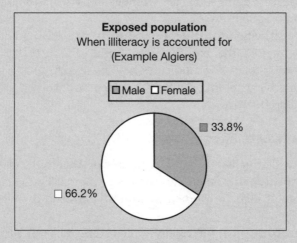

Figure 10.1.3 *Exposed population – when illiteracy is accounted for*

Similarly the following categories were analysed from a gender perspective in order to identify factors that influence gender-specific exposure to climatic changes in Algiers:

- Category 2: Influence, decision-making
- Category 3: Education – primary enrolment
- Category 4: Adult economic activity, including income differences

Aggregated exposure

The aggregated exposure of all the factors was based on all the categories, and was compiled as follows:

(Illiteracy) + (Decision-making) + (Primary Education) + (Economic Activity)
4.0 = Aggregated Exposure

Considering the scope of the UNEP study, this equation indicates the aggregate difference in exposure to climate change between men and women. Continuing with the case of Algiers, the female exposure to climate change (as compared to that of men) is as follows:

Table 10.1.2 *Women exposed to changes in climate conditions (in relation to men)*

Example Algiers	
Illiteracy rate	1.96
Decision-making	15.36
Primary enrolment	1.11
Economic activity	3.09
Total exposure	21.51/4.0 = 5.38

The total exposure is calculated according to the equation outlined above, where the number 4.0 (or 400 per cent) represents the aggregated exposure from the four different categories used. As this table indicates, the aggregated value is now 5.38. In other words, based on these calculations, women are 5.38 more vulnerable to changes in climate conditions than men in Algiers.

Conclusions

The UNEP study provides statistical evidence to highlight the significantly greater exposure that women will face to the vulnerabilities of climate change in comparison to men. In doing so, the study brings to light the stark differences in political, economic and social circumstances between women and men that result in this gender-biased situation. Furthermore, the study has demonstrated the limitations caused by scarcity of data. Availability of such data could be extremely beneficial in undertaking an in-depth analysis of climate change through a gender-lens. This would avoid the need for generalizations and other such inaccuracies. Ultimately, the UNEP study calls for the inclusion of gender-based approaches and policies related to climate change.

Notes

1   This case study is based on Ashbindu Singh, Jenny Svensson and Anushka Kalyanpur (2010) 'The state of sex-disaggregated data for assessing the Impact of Climate Change', in *Proceedings World Climate Conference 3*, Geneva, August 2009

Disclaimer

The opinions expressed in this study do not necessarily reflect the ones from the UNEP. Special thanks to Anushka Kalyanpur for her assistance in editing.

References

OECD (Organisation for Economic Co-operation and Development) (2007) *OECD Environment Working Paper Screening Study: Ranking Port Cities with High Exposure and Vulnerability to Climate Extremes. Interim Analysis: Exposure Estimates*, ENV/WKP (2007)1, OECD, Paris

UNDP (United Nations Development Programme) (2008) *Human Development Report 2007–2008: Fighting Climate Change: Human Solidarity in a Divided World*, UNDP, New York

UNESCO (United Nations Educational, Scientific and Cultural Organization) (2007) *Science, Technology and Gender*, UNESCO Publications, Paris

UNSD (United Nations Statistics Division) (2007) *United Nations Statistical Commission: Sixty Years of Leadership and Professionalism in Building the Global Statistical System 1947–2007*, UNSD, UN Publications, New York

Case 10.2

Gender and Climate Information: A Case Study from Limpopo Province, South Africa

Emma R. M. Archer

Introduction

*In Dovheni village, Limpopo Province, South Africa, a
50-year-old woman farmer takes time to describe her experience of the past season to an interviewer. Despite a fairly
reliable source of supplementary irrigation, her crops have
suffered water stress under the intense heat and inadequate
rainfall of the late season. As a result, although she has access
to customers who would like to buy her products, neither
the product yield nor the quality allow her to enter into the
long term supply contracts she plans for. Since her crop sale
income pays for her son's university fees, his enrolment for
the following academic year at the University of Venda is no
longer assured.* (Vhembe District household interview,
August 2003)

The term 'climate information system' refers to a system of
producing predictive climate information, tailoring or interpreting
the information for different sectors, disseminating the information
and applying it at the user-scale. Climate information can serve as a
tool to mitigate the negative effects of climatic risk, potentially allowing
a user to employ *ex ante* knowledge to make appropriate choices –
for example switching to a drought resistant cultivar, lower animal
stocking levels, erecting flood barriers or reinforcing roofing. Climate
information can also potentially be used to support the development
of strategies for adapting to longer-term climate change. For example,
measures taken in response to a seasonal forecast of below normal
precipitation can also serve to improve contingency planning options
in the case of increasing frequency of climatic risk over the longer

term. Providing information that can more effectively support the development of response measures that improve resilience to extreme climatic events is particularly critical given the scenarios presented in South Africa's latest Second National Communication to the United Nations Framework Convention on Climate Change (UNFCCC). In that communication higher temperatures and higher rates of evapotranspiration, as well as more intense rainfall events are predicted.

The case study

Gaps and weaknesses in the ability of climate information systems to fulfill these functions persist, however. The overall study cited here presents what was learned from more than three years of research on the usefulness of climate information in South Africa at two different scales – that of the end-user of climate information (in a resource-poor farming community in northern rural South Africa), and that of the forecast producer and key national intermediaries. Aspects of the case study presented here focus particularly on end-users of the forecast, drawn from more than two and a half years of fieldwork in Mangondi village, located in Vhembe district, Limpopo Province in northeast South Africa.

After transition to a post-apartheid government in 1994, the Northern Province (now renamed Limpopo Province) was demarcated, comprising the former Northern Transvaal province, as well as the former 'independent' ethnic homelands of Lebowa, Gazankulu and Venda (Constitution of South Africa, 1996 – Act 108). Since homeland areas under apartheid were critically underdeveloped, the province has inherited major challenges with regard to poverty and a backlog of basic services.

Mangondi community comprises approximately 400 families and is located 14km east of Thohoyandou, the former capital of the independent homeland Venda. The community experiences challenges typical of a rural former homeland area, such as limited piped domestic water in dwellings, limited communication facilities and low disposable incomes. All agriculture practiced by farmers in Mangondi community can be described as 'small-scale' or 'developing' agriculture. The study sought to identify constraints and opportunities in the application of seasonal climate forecasts amongst farmers involved in a communal food garden in Mangondi.

There were 46 farmers included in a detailed household survey in 2000 and extended through regular stakeholder and participatory interviews undertaken between 2000–2003.

As shown below, for the overall study the climate information system was investigated at two scales – that of the end-user and that of national scale forecast producers and intermediaries. Both perspectives are essential in diagnosing weaknesses and identifying strategies to improve potential benefits of the climate information system. As mentioned earlier, we focus only on the findings as they pertain to the end user in the case study presented here.

Main findings

Work at the end-user scale in Mangondi community in Limpopo Province shows that the notion of the 'end-user' in forecast applications needs to be 'sophisticated' or better characterized. As discussed elsewhere (Archer, 2003), the sample of farmers surveyed in Mangondi community is representative of the gender differentiation found in agriculture in the former homeland area in that it is mostly comprised of women. Such gender differentiation is partly a function of the tendency for male household members to seek work in metropolitan areas, while women remain in the homeland, or former homeland, areas. Findings show, however, that although women farmers are essential to food security in South Africa and, indeed, in sub-Saharan Africa (IFPRI, 2000), they may be excluded from the potential benefits of the climate information system.

Mangondi results showed that gender is significant in determining key preferences for forecast distribution (such as the medium of communication – Archer, 2003), and for determining constraints on potential uptake. For example, women were shown to have less access to credit (Archer, 2003). This would be important, for example, should a farmer require credit to purchase a drought resistant maize cultivar in response to a seasonal forecast of below normal rainfall. Women farmers were also less likely to access external sources of agricultural advice (Archer, 2003). Agricultural advisories, as shown earlier, are now being provided on the basis of the seasonal forecast, and have been shown anecdotally to have had good reception amongst the agricultural community. If women farmers, as suggested by the Mangondi data, have lower points of connectivity to external agricultural advice, they may be disadvantaged in their ability to translate forecast information into appropriate agricultural response strategies.

In short, the end-user survey to date in Mangondi community has shown that forecast utility within a community, and indeed within a household, cannot be automatically assumed and, for example, may have different utility to men and women. Gendered preferences and constraints in regard to the seasonal climate forecast, for example those described above such as women's lesser access to agricultural inputs and advice, are significant. If such preferences and constraints are not accommodated, a critical agricultural user group may not benefit from its potential in mitigating the negative effects of climatic risk (Archer, 2003).

Acknowledgements

The author would like to acknowledge the generous support of the National Oceanic and Atmospheric Administrations (NOAA) Office of Global Programs, as well as University Corporation for Atmospheric Research (UCAR) Visiting Scientist Programs for excellent logistical support. Further thanks are due to Dr Bill Easterling for his postdoctoral supervision, and to the International Research Institute for Climate & Society (IRI) for hosting. Finally, thanks are due to farmers and institutions described, who remain central to efforts to mitigate adverse effects of climatic risk in South Africa.

References

Archer, E. R. M. (2003) 'Identifying underserved end-user groups in the provision of climate information', *Bulletin of the American Meteorological Society*, vol 84, no 11, pp1525–1532

IFPRI (International Food Policy Research Institute) (2000) *Women, the Key to Food Security: Looking into the Household*, IFPRI, Washington, DC

Index